# The Properties of Water and their Role in Colloidal and Biological Systems

INTERFACE SCIENCE AND TECHNOLOGY
Series Editor: ARTHUR HUBBARD

*In this series:*

Vol. 1: Clay Surfaces: Fundamentals and Applications
Edited by F. Wypych and K.G. Satyanarayana

Vol. 2: Electrokinetics in Microfluids
By Dongqing Li

Vol. 3: Radiotracer Studies of Interfaces
Edited by G. Horányi

Vol. 4: Emulsions: Structure Stability and Interactions
Edited by D.N. Petsev

Vol. 5: Inhaled Particles
By Chiu-sen Wang

Vol. 6: Heavy Metals in the Environment: Origin, Interaction and Remediation
Edited by H.B. Bradl

Vol. 7: Activated Carbon Surfaces in Environmental Remediation
Edited by Teresa J. Bandosz

Vol. 8: Tissue Engineering: Fundamentals and Applications
By Yoshito Ikada

Vol. 9: Particles at Interfaces: Interactions, Deposition, Structure
By Zbigniew Adamczyk

Vol. 10: Interface Science in Drinking Water Treatment:
Theory and Applications
Edited by G. Newcombe and D. Dixon

Vol. 11: Surface Complexation Modelling
Edited by Johannes Lützenkirchen

Vol. 12: Theory of Colloid and Interfacial Electric Phenomena
By Hiroyuki Ohshima

Vol. 13: Sorbent Deformation
By A.V. Tvardovskiy

Vol. 14: Advanced Chemistry of Monolayers at Interfaces:
Trends in Methodology and Technology
Edited by Toyoko Imae

Vol. 15: Macromolecules in Solution and Brownian Relativity
By Stefano A. Mezzasalma

Vol. 16: The Properties of Water and their Role in Colloidal and
Biological Systems
By Carel Jan van Oss

INTERFACE SCIENCE AND TECHNOLOGY–VOLUME 16

# The Properties of Water and their Role in Colloidal and Biological Systems

Carel Jan van Oss

Department of Microbiology and Immunology
School of Medicine and Biomedical Science
State University of New York at Buffalo
South Campus, Buffalo
New York 14214-3000
USA

ELSEVIER

Amsterdam • Boston • Heidelberg • London • New York • Oxford
Paris • San Diego • San Francisco • Singapore • Sydney • Tokyo
Academic Press is an imprint of Elsevier

Academic Press is an imprint of Elsevier
Radarweg 29, PO Box 211, 1000 AE Amsterdam, The Netherlands
Linacre House, Jordan Hill, Oxford OX2 8DP, UK
32, Jamestown Road, London, NWI 7BY, UK
30 Corporate Drive, Suite 400, Burlington, MA 01803, USA
525 B Street, Suite 1900, San Diego, CA 92101-4495, USA

First edition 2008



ISBN: 978-0-12-374303-9

ISSN: 1573-4285



Printed and bound in The Netherlands

08 09 10 11 12 10 9 8 7 6 5 4 3 2 1

# Contents

# Preface

Of all liquids on this Earth, liquid water is the most pervasive and its properties are the most influential in all colloidal and, especially, in all biological systems. However, it was only since the late 1980's that it became possible to understand the non-covalent physicochemical properties of water and their influence on everything that is immersed in it, in a truly quantitative manner. The most important property of liquid water, with a major influence on all surfaces, particles, cells and molecules that are immersed in it is its very strong energy of cohesion, which is for 30% due to van der Waals attractions and for 70% a consequence of the hydrogen-bonding driven (i.e., polar) attraction between the water molecules.

This polar attraction between water molecules in liquid water causes the strong *attraction* between all hydrophobic (non-polar) molecules and particles immersed in it, known as "*the hydrophobic effect*." Conversely, the polar attraction between water molecules and hydrophilic (polar) molecules and particles, is the cause of "*hydration pressure*," i.e., the *repulsion* between such hydrophilic molecules and particles when immersed in water.

A few, out of the many examples of the influence of the properties of water on colloidal or biological systems treated in this book are:

Hydration pressure effects in water keep our blood cells from clumping together in our peripheral blood circulation and they also keep our blood serum proteins in stable solution.

In the liquid form, water occurs in clusters of about 4.5 water molecules per cluster, at room temperature. Increases in the water temperature causes a decrease in the cluster size of water, thus increasing the electron-accepticity of water, which in turn causes the aqueous solubility of most hydrophilic solutes to be greater in warm than in cold water.

The aqueous solubility of solute molecules is directly linked to their interfacial tension with water, so that such interfacial tensions, which often are difficult to measure directly, can be derived from the known aqueous solubility of these solute molecules.

Hydrophobic macro-molecules or particles clump together when immersed in water, driven by the "hydrophobic effect," which is caused by the hydrogen-bonding energy of cohesion of the water molecules that surround these hydrophobic entities. On the other hand, *single* hydrophobic atoms or small molecules, immersed in water, become individually surrounded by a sphere of water molecules, thus forming water-cages or "clathrates."

It can be demonstrated that, paradoxically, the water–air interface is the most hydrophobic (= "water-fearing) surface known to Man. For instance, "Rough" solid surfaces, when in contact with water, give the appearance of being hydrophobic because of the air trapped between the solid protrusions that are the cause of the roughness.

Hydration pressure causes the osmotic pressure of hydrophilic, electrostatically neutral, concentrated solutions of polymers (such as dextran or polyethylene glycol) to generate osmotic pressures of up to 300 atmospheres, which is hundreds of times higher than would be predicted when using the classical van 't Hoff equation that normally governs osmotic pressures.

Carel Jan van Oss
2008

## *ACKNOWLEDGEMENTS*

I am greatly obliged to Dr. John Hay (Grant T. Fisher Professor and Chairman of the Department of Microbiology and Immunology, State University of New York at Buffalo), who made it possible for me to work on this book, and to Mrs. Hillary A. Hurwitz for her expert help with the tables.

CHAPTER ONE

# General and Historical Introduction

## Contents



## Preamble

Water is the most polar[1] liquid known to Man. At room temperature (20 °C) its total free energy of cohesion, $\Delta G^{\text{cohesion}} = -145.6$ mJ/m$^2$, consisting of $\Delta G^{\text{van der Waals}} = -43.6$ mJ/m$^2$ and $\Delta G^{\text{Polar}} = -102$ mJ/m$^2$. Thus the van der

[1] There is of course mercury (Hg), which is also a liquid at room temperature and which has an apolar (van der Waals) free energy of cohesion that is about 9.2 times greater than that of water and a non-van der Waals part of the total free energy of cohesion that is about 5.6 times greater than that of water, but that non-van der Waals part is not really *polar*, as it is more accurately described as a *metallic* bond (see Chaudhury, 1987, and see also van Oss et al., 1988, 1994, p. 157).

*Interface Science and Technology*, Volume 16
ISSN 1573-4285, doi: 10.1016/S1573-4285(08)00201-9

Waals part of the free energy of cohesion of liquid water represents only 30% and the polar part represents 70% of the total. This was already known since Fowkes (1963, 1964, 1965). In addition, with respect to the interaction energies between non-polar molecules (e.g., alkanes), *when these are immersed in water*, the combining rules for apolar interactions in such cases generally causes $\Delta G^{\text{van der Waals}}$ to be rather small, leaving mainly the polar free energy of attraction of $-102$ mJ/m$^2$, which thus represents close to 100% of the total free energy of interaction in water among non-polar molecules or particles.

As was only realized much later, these $-102$ mJ/m$^2$, representing the polar (in this case the hydrogen-bonding) free energy of cohesion of water, also happen to be the sole driving force for the *hydrophobic effect*. Notwithstanding these new data and probably mainly due to a continuing indecisiveness as to which forces were apolar and which were polar (see Sub-section 2.3, below) no significant advances were made in this matter for about another 20 years after Fowkes (1963, 1964, 1965). Finally, based on important clarifications proposed by Chaudhury (1984) and starting in early 1985, Chaudhury and I began (mainly *via* long-distance telephone) to develop the combining rules which allow the quantitative expression of polar free energies in SI units (van Oss, Chaudhury and Good, 1987, 1988). The ensuing results allowed the polar free energies of interaction to be combined with the van der Waals interaction energies (and the electrical double layer interaction energies where applicable), into a complete system comprising all non-covalent interactions taking place in and with liquid water (see also van Oss, 1994, 2006).

Now, more than another two decades past 1987, this book aims to treat the combined non-covalent non-polar, polar and electrical double layer interactions taking place in and with water, from the viewpoint of all the germane physical and physico-chemical properties of liquid water.

## 1. Some Examples of Polar Forces Interacting in the Mammalian Blood Circulation

Essentially all repulsive as well as attractive non-covalent interactions at a colloidal scale occurring in biological systems take place in water. Some examples of such interactions in water, looking for instance at the mammalian peripheral blood circulation, include:

1. Repulsion:

   The mutual, non-specific repulsion between protein molecules, which keeps them dissolved in blood serum and permits them to avoid precipitation.

   The mutual, non-specific repulsion between leukocytes, platelets, etc., which keeps them in stable suspension in the blood and lymph circulation and thus prevents the formation of thrombi.

   (Thus the principal constituents of blood can safely circulate in their aqueous environment.)

2. Attraction:

The specific attraction between pseudopodia of phagocytic leukocytes and bacteria that have found their way into the bloodstream allows the phagocytic cells to internalize such bacteria and destroy them.

The specific attraction between the epitope of a (foreign) antigen and the paratope of a complementary antibody molecule (immunoglobulin) triggers a series of events leading to the foreign antigen's destruction, see also Chapter 4, Section 6.

Whilst electrical double layer forces sometimes play a role in both the specific attractions and the non-specific repulsions alluded to above, polar, Lewis acid–base (AB) forces tend to accompany such electrical double layer forces and are often stronger than these, especially in high ionic strength media, such as the human blood circulation, in which electrostatic interactions are significantly attenuated. However, polar AB forces also act quite well in the absence of any electrical double layer forces, see Chapter 4, Sections 2, 3 and Chapter 7, Section 4, below.

## 2. EARLY EXAMPLES OF THE TREATMENT OF NON-COVALENT INTERACTIONS IN WATER

### 2.1 DLVO and Non-DLVO forces

In the 1940's the interaction energies between particles, immersed in a liquid were assumed to obey what is now known as the DLVO theory, after Derjaguin and Landau (1941), from the USSR and Verwey and Overbeek (1948), from the Netherlands; however, see also Hamaker (1936, 1937a, 1937b, 1937c). The classical version of this approach takes into account van der Waals attractive and electrostatic repulsive energies, acting upon such particles, as a function of interparticle distance. The classical DLVO theory does not account for polar (e.g., hydrogen-bonding) interactions. It does turn out, however that, *in water*, its polar (largely hydrogen-bonding) free energies of adhesion (to other entities) as well as of cohesion (within itself), usually account for approximately 90% of all polar, i.e., non-covalent interaction energies that are active in that liquid. [Even though van der Waals forces represent 30% of the cohesive interaction energies of water (see above), the van der Waals combining rules for interactions *in liquids* are such that van der Waals interaction energies among most biological compounds or particles, *immersed in water*, usually account for at most only a few % of the total, due to the van der Waals part of the combining rules given in Chapter 5, Eq. (5.1).]

It was only in the 1980's that hard experimental evidence, obtained with Israelachvili's force balance (Israelachvili, 1985, 1991), which permits the direct measurement of interaction forces in water, revealed the existence of *"non-DLVO" forces* which are neither due to van der Waals nor to electrical double layer forces (see also Claesson, 1986, and a more recent review article by Grasso et al., 2002).

## 2.2 Good's introduction of a Φ-factor and Fowkes' evaluation of a van der Waals/Non-van der Waals ratio of the surface tension of water

Meanwhile, from a theoretical point of view, in 1957 Good (see Girifalco and Good, 1957) had remarked upon the very different interfacial behavior of organic molecules when immersed in water, as compared to their interfacial behavior when immersed in various organic liquids. Thus Girifalco and Good (1957) proposed the use of a factor, $\Phi$, of which the degree of departure from unity could be used as an indicator of the polarity of the system. It was the value of the (measured) interfacial tensions between oils and water (Girifalco and Good, 1957; see also Harkins, 1952), which we now know to be only valid for alkane–water systems and not for polar oil–water systems (see Chapter 3) and which for alkanes only reached a maximum value of around 51 mJ/m$^2$, that allowed Fowkes (1963, 1964, 1965), using alkane–water interfacial tension values, to deduce the ratio of van der Waals to non-van der Waals contributions to the surface tension of water at 20 °C as being equal to 21.8/51.0 mJ/m$^2$, i.e., as 30/70%. Later, using direct contact angle measurements with apolar liquids on aqueous gels of different concentrations, van Oss, Ju et al. (1989) confirmed the correctness of Fowkes' findings for the non-polar (van der Waals) component of the surface tension of water.

## 2.3 The three van der Waals forces: Are some of them polar?

It took, however, until 1985 before it became possible to begin to delineate with some clarity what the real difference was between apolar and polar forces in water. A remaining difficulty between knowing that there was a problem and solving it was that there are three different and well-known parts to the electrodynamic forces, together representing the sum-total of van der Waals forces. These are, in the order of their discovery:

I. Van der Waals–Keesom, or dipole–dipole (orientation) forces (Keesom, 1915, 1920, 1921).
II. Van der Waals–Debye, or dipole–induced dipole (induction) forces (Debye, 1920, 1921).
III. Van der Waals–London, or fluctuating dipole–induced dipole (dispersion) forces (London, 1930).

Quantitatively the London or dispersion forces are by far the most important ones of the three and Fowkes made use of that fact and simply equated dispersion forces with the whole of the van der Waals forces, which is how he got the right answer for the proportion of van der Waals contributions to the polar cohesion of water as early as 1963. Meanwhile for most scientists seeking a clear distinction between non-polar and polar interactions, the semantic distinction between these two even among the three types of van der Waals forces was by no means clear. Many had an inkling that Fowkes (1963, 1964, 1965) could be right and that the London (dispersion) forces most likely represented the non-polar ones, but the dipole–dipole (Keesom) and the dipole–induced dipole (Debye) forces still sounded very polar.

## 3. Macroscopic-Scale Interactions, Chaudhury's Thesis and Lifshitz–van der Waals Forces

On a microscopic level all three varieties of van der Waals energies between, e.g., atoms or small molecules, are known to decay very steeply as a function of distance (d), as $d^{-6}$, although it should be noted that in the 1930's, when Hamaker (1936, 1937a) wrote two papers which can be seen as precursors to the DLVO theory, he, as well as Verwey and Overbeek (1948) in the 1940's considered only van der Waals–London interactions in their publications on decay vs distance considerations comprising van der Waals energies as the attractive component. Meanwhile, Hamaker (1937b), using a pair-wise additivity approach, also preceded Lifshitz (1955) in proposing (correctly) that *on a macroscopic level* (non-retarded) van der Waals interactions between larger spherical molecules and other spherical bodies decay as a function of inter-sphere distances as $d^{-1}$; see Chapter 7, Section 2, below. [Lifshitz (1955) improved on Hamaker's (1937b) pair-wise additivity method by using a more rigorous approach, involving dielectric constants; see especially Visser (1972), who also demonstrated the complexities involved in arriving at really accurate Hamaker constants.] In this book we shall consider only van der Waals interaction energies on a macroscopic scale of at least medium-sized molecules (e.g., hexane, benzene, etc.), as well as of larger quasi-spherical molecules, particles or cells, when in contact with water, which obey the $d^{-1}$ law. We shall henceforth allude to these interactions as Lifshitz–van der Waals (LW) interactions, see below.

In December 1984 Chaudhury defended his thesis on the surface-thermodynamic properties of larger entities, based on Lifshitz's theory (1955). He showed that on a macroscopic scale as defined above, *all three varieties* of van der Waals energies decay in the same manner as a function of distance (i.e., as $d^{-1}$ for two spheres and for a sphere and a flat plate, and as $d^{-2}$ for two flat parallel plates) and that what we now call Lifshitz–van der Waals (LW) energies (van Oss, Chaudhury and Good, 1988) should be considered *non-polar*. Thus, as of 1985 the way appeared clear to identify the *polar* forces operating in and with water as distinct from LW forces.

## 4. Rules for Repulsive Apolar (van der Waals) Forces between Different Polymers Dissolved in an Apolar Liquid, Compared with the Rules for Repulsive Polar (Lewis Acid–Base) Forces between Identical Polymers, Particles or Cells, Immersed in Water

### 4.1 Van der Waals repulsions between different materials immersed in an apolar liquid

Everybody knows that van der Waals forces are always attractive. *In vacuo* or in air this is indeed true, but in liquids it is only true for van der Waals forces acting on *identical* particles or molecules, immersed in the liquid. Hamaker (1937b) already mentioned that there can be circumstances in which two *different* non-polar materials, immersed in a liquid medium could repel one another via van der Waals–London forces, whilst two *identical* non-polar molecules or materials, when immersed in a liquid, always had to attract each other. Hamaker even hinted at the experimental conditions under which such a van der Waals repulsion might be possible. Derjaguin (1954) suggested that van der Waals repulsions could exist, but did not elaborate upon the conditions that would favor such a repulsion. Finally, Visser (1972) explained precisely under which conditions van der Waals repulsions should occur, which is when two *different* solutes, 1 and 2, are immersed in a liquid, 3, where the Hamaker constant of solute 1 is greater than the Hamaker constant of the liquid, 3 and the value of the Hamaker constant of solute 2 is smaller than that of the liquid (or vice-versa). [The Hamaker constant A, is a constant directly related to the dielectric permeability properties of a given material, see van Oss, Chaudhury and Good (1988) but A is also directly proportional to the Lifshitz–van der Waals surface tension component of that material; see Chapter 2, Section 3, below]. In 1967 Fowkes suggested different materials that might undergo such a van der Waals repulsion (Fowkes, 1967) and finally van Oss, Omenyi and Neumann (1979) experimentally demonstrated the validity of Visser's theory of van der Waals repulsions between two different polymers which are both dissolved in the same liquid, where a repulsion between the two kinds of dissolved polymer molecules manifests itself as a phase separation, whilst a lack of repulsion (or "compatibility") resulted in the dissolution of both polymers while remaining mixed together in the same liquid; see also van Oss, Chaudhury and Good (1989).

The major point to observe here about interactions in *non-polar systems* is the fact that it is only when two *different* compounds (1 and 2), each of which can be *characterized by one given (electrodynamic) property* (in this case expressed by their Hamaker constants, A), are dissolved in a (third) liquid (3), that 1 and 2 can repel one another when:

$$A_1 > A_3 > A_2,$$

or when:

$$A_1 < A_3 < A_2.$$

However, when apolar materials or compounds 1 and 2 are *identical*, such a repulsion cannot occur.

## 4.2 Lewis acid–base repulsions between *identical* polar materials, immersed in water

Contrary to non-polar, LW systems, it is an experimental fact that *in polar and especially in aqueous systems*, many biopolymers, particles or cells, each within their category of one and the same kind, mutually repel one another, even though any $\zeta$-potential they may have is insufficient to cause a repulsion that is stronger than the Lifshitz–van der Waals attraction. This is for instance the case with immunoglobulins dissolved in mammalian blood sera, as well as in the case of all varieties of leukocytes, freely suspended in mammalian blood; see, e.g., the examples given above in Section 1. Therefore any theory of polar interactions in water has to explain how *polar forces* between *identical entities*, immersed in water, can cause a mutual repulsion. This is one reason for being compelled to look for *dual polar properties* attached to polar molecules, because with *two different polar properties* on each single polar molecule, particle or cell, immersed in water, one can theoretically achieve a mutual polar repulsion between such identical entities, if property 1 of such a polar entity is larger than the corresponding property of water, whilst property 2 of that polar entity is smaller than the corresponding property of water. For this reason, as well as for the fact that the salient polar property of water is that of being strongly subject to self-hydrogen bonding, which in the most general sense consists of electron-accepting and electron-donating or Lewis acid–base (AB) interactions (Lewis, 1923), such AB interactions clearly impose themselves as the one mechanism fulfilling all the requirements for attractive as well as repulsive polar interaction systems in water.

Thus, the total surface tension ($\gamma$) of a polar condensed-phase material consists of:

$$\gamma = \gamma^{\mathrm{LW}} + \gamma^{\mathrm{AB}}, \tag{1.1}$$

where:

$$\gamma^{\mathrm{AB}} = 2\sqrt{(\gamma^{+}\cdot\gamma^{-})} \tag{1.2}$$

and where $\gamma^{+}$ is the electron-acceptor parameter of the surface tension of the material, and $\gamma^{-}$ is its electron-donor parameter; see van Oss, Chaudhury and Good (1988) and Chapter 2, Section 1, below. For the experimental validation of this approach (and its subsequent results), experimentally tested under both repulsive and attractive conditions, see Chapter 7, Section 2.

## 5. The Fallacy of Designating Only One Single Component to Represent the Polar Properties of the Surface Tension of a Polar Condensed-Phase Material

Meanwhile, since 1969 various authors, from Owens and Wendt (1969), Kaelble (1970), Hamilton (1974), Andrade et al. (1979), to Janczuk et al. (1990) preferred a single, polar property, identified by a subscript, as in "$\gamma^{\mathrm{P}}$" and treated exactly as $\gamma^{\mathrm{LW}}$, which was simply subject to a geometric mean treatment for obtaining the non-polar component of the interfacial tension with water (see also Neumann et al., 1974, who also used a single polar property, described in a so-called "equation of state").

It is easily shown however that a single property for $\gamma^{\mathrm{P}}$, when treated exactly like $\gamma^{\mathrm{LW}}$, results in an impossibility for the polar interfacial tension with water to assume a negative value, which is tantamount to decreeing that polar, water soluble polymers such as dextran, cannot be soluble in water (van Oss and Good, 1992). This is because the polar interfacial tension between material, i, and water, w, would then have to be expressed as:

$$\gamma^{P}_{iw} = (\sqrt{\gamma^{P}_{i}} - \sqrt{\gamma^{P}_{w}})^2 \qquad (1.3)$$

of which the right-hand side clearly cannot be negative.

In reality, two identical polar entities, i, immersed in water, w, will repel one another when:

$$\text{(A)} \quad \gamma_i^+ < \gamma_w^+ \quad \text{and} \quad \gamma_i^- > \gamma_w^-$$

or when:

$$\text{(B)} \quad \gamma_i^+ > \gamma_w^+ \quad \text{and} \quad \gamma_i^- < \gamma_w^-,$$

where contingency (A) occurs by far the most frequently (see van Oss et al., 1997).

## 6. More Recent Developments

Although many more new developments are described in subsequent chapters, in this chapter a few of the more novel as well as a few of the lesser known salient phenomena are briefly reviewed in this section.

### 6.1 Properties of water

One of the more recent developments is the growing interest in the curious properties of the water–air interface. These are for an important part due to the unusually high apparent hydrophobicity of rough surfaces, when measured via contact angle measurements with drops of water (Busscher et al., 1992). This is not so much a consequence of the roughness *sensu stricto* of these surfaces, but rather of the fact that drops of water deposited upon such rough surfaces mainly touch layers of air, enclosed in the concavities between the protruding peaks (van Oss et al., 2005). In fact, the air side of the water–air interface is the only *hydrophobic* interface in the truest sense of the word, in that it is actually practically 100% water repellent, as on a virtually ideal rough surface. Onda et al. (1996) reached a contact angle of 174° with a water drop, which is only 6° removed from the ideal, but in real life unattainable 180°. Cassie (1948) and Cassie and Baxter (1944) who worked on the water-repellency of raincoat textiles more than half a century ago, proposed a useful equation for calculating the proportion of a rough surface area that is occupied by air (see Chapter 8, Section 1).

The influence of an increase in temperature (T) on the cluster size of water (which is about 4.5 water molecules per cluster at 20 °C) concomitantly strongly influences the $\gamma^+/\gamma^-$ ratio of water. As most polar materials are preponderantly electron-donors, i.e., they have a significant $\gamma^-$ value and as the $\gamma^+$ value of water is the one that increases with an increase in T, the result is that polar entities, immersed in water, become more hydrated with an increase in T. This accounts for the increase in aqueous solubility with T, in the case of polar compounds and for the greater efficiency one usually observes when washing partly polar materials in warm rather than in cold water (see Chapter 5).

The reason why an increase in T which causes a decrease in the cluster size of water, is accompanied by an increase in the $\gamma^+/\gamma^-$ ratio of water is that the smaller the clusters become, the more H atoms, as compared to O atoms, are exposed per water cluster. The cluster size of water is also linked to its viscosity: the smaller the cluster size, the lower the viscosity of water; it is of course well known that the viscosity of water decreases sharply with T; see Chapter 9.

## 6.2 Influence of immersion in water on the behavior of non-polar and polar entities

The elucidation of the mechanisms of hydrophobic attraction as well as of hydrophilic repulsion of materials or molecules immersed in water has occurred only relatively recently. The mechanisms of these two phenomena are best understood by using the equation defining the free energy of interaction between two similar materials, immersed in water; see Chapter 5, Eq. (5.1).

The hydrophobic attraction between molecules or particles immersed in water is caused solely by the AB free energy of cohesion among the water molecules. This attraction is always present, regardless of the hydrophobicity or hydrophilicity of the immersed molecules or particles. However, the underlying hydrophobic attraction can be overcome and a net hydrophilic repulsion can be achieved when such immersed molecules or particles are very hydrophilic and thus become so strongly hydrated that they cause a net hydrophilic repulsion (sometimes called "hydration pressure") when their repulsive energy becomes stronger than the omni-present attractive energy of the hydrophobic effect in water; see Chapter 5. Thus, both hydrophobic attraction in water (the "hydrophobic effect") and hydrophilic repulsion in water ("hydration pressure") are caused by Lewis acid–base forces.

SECTION A

# Non-Covalent Energies of Interaction—Equations and Combining Rules

CHAPTER TWO

# The Apolar and Polar Properties of Liquid Water and Other Condensed-Phase Materials

## Contents



## 1. The $\gamma^{LW}$ and $\gamma^{AB}$ Equations

### 1.1 Apolar surface tensions

The surface tension of water as well as of other polar liquids, designated as $\gamma$, has an apolar plus a polar part. The first, apolar part is composed of van der Waals interactions (comprising all three van der Waals energies mentioned in Chapter 1) and grouped together as one macroscopic-scale van der Waals energy, as described by Lifshitz (1955), and which are henceforth alluded to as Lifshitz–van der Waals (LW) energies; see also Chaudhury (1984) and van Oss, Chaudhury and Good (1988). In surface tension symbolism these are designated as $\gamma^{LW}$. The difference between surface *tensions* ($\gamma$) and surface *free energies* ($\Delta G$), for instance for Lifshitz–

*Interface Science and Technology*, Volume 16
ISSN 1573-4285, doi: 10.1016/S1573-4285(08)00202-0

van der Waals interactions, is expressed as follows:

$$\Delta G^{LW} = -2\gamma^{LW}, \quad (2.1)$$

where $\Delta G^{LW}$ is the LW free energy (of cohesion) between, e.g., the molecules of liquid water and $\gamma^{LW}$ is the LW surface *tension* of (in this case) liquid water. It is therefore not advisable to call $\gamma$ (e.g., of water) the "surface *energy*" of water, as that clearly differs by a factor $-2$ from the surface *tension* of water. Nonetheless this confusing practice still is commonly encountered in the colloid and surface science literature even though it easily leads one to arrive at numerically incorrect conclusions.

Finally it should not be forgotten that completely non-polar liquids (such as alkanes) as well as non-polar solids (e.g., Teflon, polyethylene, polypropylene) have only the LW surface tension component, so that in such cases, $\gamma = \gamma^{LW}$.

## 1.2 Surface tensions of polar materials

The surface tensions of *polar* liquids (such as water) and other condensed-phase materials *all* consist of a non-polar surface tension component ($\gamma^{LW}$) *plus* a polar surface tension component, designated as $\gamma^{AB}$ (AB for Lewis acid–base); see van Oss, Chaudhury and Good (1988). Thus for *polar* liquids and other *polar* condensed-phase materials:

$$\gamma = \gamma^{LW} + \gamma^{AB} \quad (2.2)$$

(see also Chapter 1, Eq. (1.1)). $\gamma^{LW}$ has only one (apolar) surface property, linked to the material's Hamaker constant, which in turn is linked to the material's dielectric constant; see Sub-section 1.3 of Chapter 3, below. $\gamma^{AB}$ on the other hand comprises both the polar Lewis acid and the polar Lewis base surface properties of the material, i.e., its electron-accepticity ($\gamma^{+}$) and its electron donicity ($\gamma^{-}$), where each of these can vary independently from one material to the other, according to:

$$\gamma^{AB} = 2\sqrt{(\gamma^{+}\cdot\gamma^{-})} \quad (2.3)$$

so that (for all polar condensed-phase materials or compounds), combining Eqs. (2.2) and (2.3):

$$\gamma = \gamma^{LW} + 2\sqrt{(\gamma^{+}\cdot\gamma^{-})}. \quad (2.2A)$$

## 1.3 Surface and interfacial tensions

Both surface tensions and interfacial tensions are symbolized by the lower-case Greek letter $\gamma$ (gamma).

*Surface tensions* of a liquid (L) are related to the free energy of cohesion ($\Delta G^{coh}$) between the *identical* molecules of the liquid as:

$$\Delta G_{LL}{}^{coh} = -2\gamma_L, \quad (2.4)$$

where $\Delta G^{coh}$ is labeled with a double subscript (e.g., LL or SS), whilst $\gamma$ is labeled with a *single subscript* (e.g., L or S). Equation (2.4) also applies to a certain extent to solids but whilst for liquids Eq. (2.4) pertains to the *total* free energy of cohesion of liquid, L, for solids it only applies to the *non-covalent* (or, with metals to the non-metallic) part of the free energy of cohesion, which is usually much smaller than the covalent (or metallic) part. In any case, for solids (S) the non-covalent part of their cohesion is also described as:

$$\Delta G_{SS}^{\text{coh(non-covalent)}} = -2\gamma_S. \tag{2.4A}$$

Surface tensions ($\gamma_L$ or $\gamma_S$) thus are properties related to all condensed-phase materials (liquids as well as solids).

*Interfacial tensions* are tensions related to the adjoining interfaces between *two different* condensed-phase materials and are symbolized with a $\gamma$ and *two different subscripts*, e.g.: LS, or $L_1L_2$, or $S_1S_2$. The interfacial tension between two non-polar condensed-phase materials, such as a non-polar liquid and a non-polar solid is expressed as:

$$\gamma_{LS}^{LW} = (\sqrt{\gamma_L^{LW}} - \sqrt{\gamma_S^{LW}})^2. \tag{2.5}$$

This implies that the interfacial tension between two *apolar* materials, L and S, can never be negative. For *polar* materials:

$$\begin{aligned}\gamma_{LS} &= \gamma_{LS}^{LW} + \gamma_{LS}^{AB} \\ &= (\sqrt{\gamma_L^{LW}} - \sqrt{\gamma_S^{LW}})^2 + 2[\sqrt{(\gamma_L^+\gamma_L^-)} + \sqrt{(\gamma_S^+\gamma_S^-)} \\ &\quad - \sqrt{(\gamma_L^+\gamma_S^-)} - \sqrt{(\gamma_L^-\gamma_S^+)}],\end{aligned} \tag{2.6}$$

where, as can be seen in the polar part of the equation, the electron-acceptor of S is allowed to interact with the electron-donor of S as well as with the electron-donor of L and *vice-versa*, hence the four different parts of the $\gamma_{LS}^{AB}$ section of Eq. (2.6). The $\gamma_{LS}^{AB}$ section of Eq. (2.6) shows its first two terms as positive (here indicating cohesion) and the next (last) two terms as negative (indicating adhesion), so that all polar combinations are represented. The four-term $\gamma_{LS}^{AB}$ part of Eq. (2.6) can also be written as:

$$\gamma_{LS}^{AB} = 2(\sqrt{\gamma_L^+} - \sqrt{\gamma_S^+})(\sqrt{\gamma_L^-} - \sqrt{\gamma_S^-}) \tag{2.7}$$

from which follows that $\gamma_{LS}^{AB}$ can have a *negative value*, when:

$$\gamma_L^+ > \gamma_S^+ \quad \textit{and} \quad \gamma_L^- < \gamma_S^- \tag{\textbf{A}}$$

or when:

$$\gamma_L^+ < \gamma_S^+ \quad \textit{and} \quad \gamma_L^- > \gamma_S^-, \tag{\textbf{B}}$$

where contingency (**A**) is by far the most common one when the liquid, L, is *water*; see also Chapter 1, Section 5.

With the exception of Eq. (2.1), which is the simplest member of the Dupré equation family, all its other versions comprise *interfacial* as well as *surface* tensions for the expression of the various forms of $\Delta G$; see the following sub-section. In surface thermodynamics all of these: $\gamma_1$ (surface tensions), $\gamma_{12}$ (interfacial tensions) and all modes of $\Delta G$ (free energies of interaction) are usually quantitatively expressed in (S.I.) units of $mJ/m^2$.

## 1.4 The Dupré equations

What one might call the original Dupré equation for the free energy of interaction between condensed-phase materials, 1 and 2 ($\Delta G_{12}$), *in vacuo* or in air, reads as follows:

$$\Delta G_{12} = \gamma_{12} - \gamma_1 - \gamma_2. \tag{2.8}$$

Most authors cite Anastase Dupré (1869) as the originator of Eq. (2.8). One would however search his "Théorie Mécanique de la Chaleur" of 1869 in vain for the expression of the correct form of this equation. An erroneous form of Eq. (2.8) (featuring $2\gamma_{12}$ instead of $\gamma_{12}$) *does* appear in the book and our guess is that a student or a successor of Dupré's applied the necessary correction in later quotations of the work. Meanwhile, Eq. (2.8) as well as its variants, discussed below, continues to be alluded to as the Dupré equation. It can also be shown that Eq. (2.4) can be derived from Eq. (2.8) (which is the basic and original Dupré equation), for the case where $\gamma_1$ has the same value as $\gamma_2$, so that then:

$\Delta G_{11} = -2\gamma_1$, because the *interfacial tension* between two identical materials (1 and 1) is always zero (also noting that $\gamma_1$ is always positive).

Equation (2.8) can be written in its fully polar form, incorporating Eqs. (2.6), (2.2) and (2.3):

$$\Delta G_{12} = \gamma_{12} - \gamma_1 - \gamma_2 = -2[\sqrt{(\gamma_1^{LW}\gamma_2^{LW})} + \sqrt{(\gamma_1^{+}\gamma_2^{-})} + \sqrt{(\gamma_1^{-}\gamma_2^{+})}\,], \tag{2.9}$$

which shows that $\Delta G_{12}$ is always negative, i.e., attractive.

Equation (2.4), in its fully polar form is (Eqs. (2.1)–(2.3)):

$$\Delta G_{11} = -2\gamma_1 = -2\gamma_1^{LW} - 4\sqrt{(\gamma_1^{+}.\gamma_1^{-})}. \tag{2.10}$$

Then there is the case of the free energy of interaction between two *identical* solid materials, S, *immersed in liquid*, L, where:

$$\begin{aligned}\Delta G_{SLS} &= -2\gamma_{SL} = -2\gamma_{12}\\ &= -2(\sqrt{\gamma_1^{LW}} - \sqrt{\gamma_2^{LW}})^2 - 4[\sqrt{(\gamma_1^{+}\gamma_1^{-})} + \sqrt{(\gamma_2^{+}\gamma_2^{-})}\\ &\quad - \sqrt{(\gamma_1^{+}\gamma_2^{-})} - \sqrt{(\gamma_1^{-}\gamma_2^{+})}\,].\end{aligned} \tag{2.11}$$

Equation (2.11) is the object of closer scrutiny in Chapter 5 and is also often used in Chapters 7 and 8.

Finally, the Dupré equation pertaining to the free energy of interaction between two *different* condensed-phase materials, 1 and 2, immersed in a liquid, 3, is expressed as:

$$\Delta G_{132} = \gamma_{12} - \gamma_{13} - \gamma_{23}, \tag{2.12}$$

which, expanded into its polar version becomes:

$$\begin{aligned}\Delta G_{132} = 2[&\sqrt{(\gamma_1^{LW}\gamma_3^{LW})} + \sqrt{(\gamma_2^{LW}\gamma_3^{LW})} - \sqrt{(\gamma_1^{LW}\gamma_2^{LW})} - \gamma_3^{LW}\\ &+ \sqrt{\gamma_3^{+}}(\sqrt{\gamma_1^{-}} + \sqrt{\gamma_2^{-}} - \sqrt{\gamma_3^{-}}) + \sqrt{\gamma_3^{-}}(\sqrt{\gamma_1^{+}} + \sqrt{\gamma_2^{+}} - \sqrt{\gamma_3^{+}})\\ &- \sqrt{(\gamma_1^{+}\gamma_2^{-})} - \sqrt{(\gamma_1^{-}\gamma_2^{+})}\,].\end{aligned} \tag{2.13}$$

Equation (2.13) is especially germane to Chapters 6, 12 and 14.

## 1.5 The Young equation

Even though Thomas Young's (1805) "Essay" contained only English prose and was devoid of any formulas or equations, what we now know as the Young equation could be distilled from the text and is generally expressed as follows:

$$\gamma_L \cos\theta = \gamma_S - \gamma_{SL}. \qquad (2.14)$$

Thus Dr. Thos. Young was probably the first to define the concept of interfacial tension ($\gamma_{SL}$). Surface and interfacial tensions have been introduced before (see Eqs. (2.4), (2.4A) and (2.6), above) whilst the contact angle in degrees measured with liquid, L, deposited on a flat surface of the solid, S is given in Sub-section 1.6.1, below. As $\gamma_S$, $\gamma_L$ and $\theta$ are relatively easily measured, Young's equation plus contact angle measurements would present an easy method for finding the interfacial tension, $\gamma_{SL}$, between S and L. To a certain extent that is indeed the case but it should not be overlooked that in polar systems both $\gamma_S$ and $\gamma_L$ are more complex than would at first sight appear; see, e.g., Eqs. (2.2), (2.3) and (2.6), above. Fortunately a solution to this complication can be found in the use of the Young–Dupré equation; see the following Sub-section 1.6, below.

## 1.6 The Young–Dupré equation and contact angle determination

### 1.6.1 The Young–Dupré equation

When one combines the Dupré equation (Eq. (2.8)) written as:

$$\Delta G_{SL} + \gamma_L = \gamma_{SL} - \gamma_S$$

with Young's equation (Eq. (2.14)), written as:

$$-\gamma_L \cos\theta = \gamma_{SL} - \gamma_S$$

one obtains:

$$\gamma_L + \gamma_L \cos\theta = -\Delta G_{SL}$$

or:

$$(1 + \cos\theta)\gamma_L = -\Delta G_{SL} = -\Delta G_{SL}{}^{LW} - \Delta G_{SL}{}^{AB} \qquad (2.15A)$$

or expanded to the non-polar *plus* the polar form:

$$(1 + \cos\theta)\gamma_L = 2[\sqrt{(\gamma_S{}^{LW}\gamma_L{}^{LW})} + \sqrt{(\gamma_S{}^{+}\gamma_L{}^{-})} + \sqrt{(\gamma_S{}^{-}\gamma_L{}^{+})}\,], \qquad (2.15B)$$

which is the Young–Dupré equation as used for polar systems. The contact angle, $\theta$, is easily measured and $\gamma_L$, for many liquids, at various temperatures, can usually be found in Jasper's (1972) tables, or else is also easily measured, e.g., with a Wilhelmy plate, or via the shape of a hanging drop in air (Adamson, 1990). However, it should not be forgotten that, for polar surfaces, one has three unknowns, i.e., $\gamma_S{}^{LW}$, $\gamma_S{}^{+}$ and $\gamma_S{}^{-}$, so that on any given surface, S, one must do at least three contact angle determinations, with three different liquids, to solve for all three unknowns. It is usually best to use diiodomethane to determine $\gamma_S{}^{LW}$, plus two polar liquids, of which one should be water. See Table 2.1 for the $\gamma$ values for the various contact angle liquids; see also Figure 2.1 and the following sub-section.

**Table 2.1** $\gamma$, as well as $\Delta G_{LWL}$ values for various apolar and polar liquids, at 20 °C, in mJ/m²

| Liquids | $\gamma_L$ | $\gamma^{LW}$ | $\gamma^+$ | $\gamma^-$ | $\gamma^{AB}$ | $\Delta G_{LWL}$ |
|---|---|---|---|---|---|---|
| Octane[a] | 21.6 | 21.6 | 0 | 0 | 0 | −102.0 |
| Decane[a] | 23.8 | 23.8 | 0 | 0 | 0 | −103.1 |
| Dodecane[a] | 25.4 | 25.4 | 0 | 0 | 0 | −102.3 |
| Tetradecane[a] | 26.6 | 26.6 | 0 | 0 | 0 | −102.5 |
| Hexadecane[e] | 27.5 | 27.5 | 0 | 0 | 0 | −102.7 |
| Diiodomethane[b] | 50.8 | 50.8 | ≈0.01 | 0 | 0 | −112.1 |
| Water[b] | 72.8 | 21.8 | 25.5 | 25.5 | 51.0 | NA[g] |
| Glycerol[b] | 64.0 | 34.0 | 3.92 | 57.4 | 30.0 | +28.3 |
| Formamide[b] | 58.0 | 39.0 | 2.28 | 39.6 | 19.0 | +12.6 |
| Ethylene glycol | 48.0 | 29.0 | 3.0 | 30.1 | 19.0 | +2.4 |
| Chloroform[c,f] | 27.3 | 27.3 | 1.5 | 0 | 0 | −77.9 |
| Benzene[d] | 28.85 | 28.85 | 0 | 0.96 | 0 | −83.2 |
| Ethyl ether[c,f] | 17.0 | 17.0 | 0 | 9.0 | 0 | −42.0 |
| Ethyl acetate[c,f] | 23.9 | 23.9 | 0 | 6.2 | 0 | −51.8 |

*a* Can be used to determine $R_a$ (Eq. (2.18)) used in wicking.

*b* Used in direct contact angle measurements as well as in wicking.

*c* Data from van Oss, Wu et al. (2001).

*d* Data from van Oss (2006, Chapter XVII); van Oss, Docoslis and Giese (2001, 2002); van Oss, Giese and Docoslis (2001); van Oss et al. (2001); van Oss, Giese and Good (2002).

*e* Used in Microbial Adhesion to Hydrocarbons (MATH), see Rosenberg et al. (1980).

*f* Used in Microbial Adhesion to Solvents (MATS). Bellon-Fontaine et al. (1996) and Meylheuc et al. (2001).

*g* NA = not applicable.

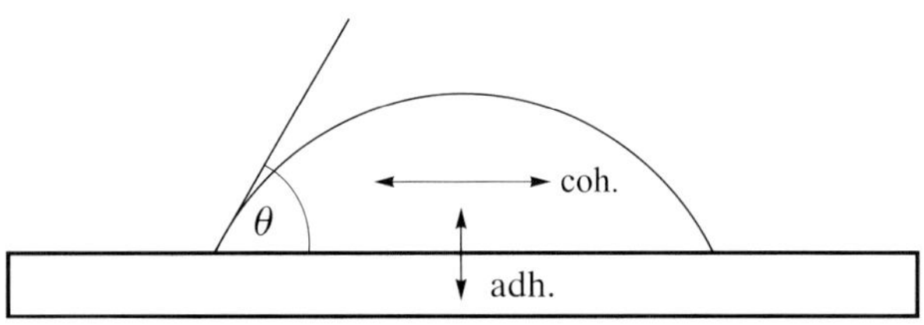


**Figure 2.1** Graphic depiction of a contact angle of liquid, L, deposited upon a flat, smooth, horizontal surface of a solid, S. The contact angle, $\theta$, is always measured through the drop, at the tangent to the drop at the triple point: solid–liquid–air. For further explanation see text [Sub-section 1.6.2; see also van Oss (1994, 2006)].

## 1.6.2 The contact angle as a force balance

Figure 2.1 illustrates how the sessile drop of a liquid, L, which forms a contact angle when deposited on a flat horizontal surface of a solid, S, can serve as a force balance by means of its shape, which is completely determined by the interplay between the free energy of *cohesion* of a liquid, L, which is exactly equal to the free energy of *adhesion* between liquid, L and solid, S, at equilibrium. In Figure 2.1 the free energy of cohesion of liquid L of the drop is indicated by a double-headed horizontal arrow, whilst the free energy of adhesion between L and S is indicated by a double-headed vertical arrow, of equal length.

### 1.6.3 Direct contact angle measurement

Whilst it is of course only the sessile drop which can represent the equilibrium situation depicted in Figure 2.1, in actual fact one should attempt to measure the "advancing" contact angle ($\gamma_a$), i.e. one should let the drop advance by slowly feeding more of the liquid through a vertical syringe placed above the drop, with the needle barely entering the top of the drop, and then measure immediately at the very moment the drop stopped advancing (Good, 1979). This is to assure that the circular line around the drop, where it touches the solid surface, i.e., at the solid–liquid–air line, will only touch as yet unwetted solid at the air side.

There are, however, some authors (see, e.g., Fowkes et al., 1980) who believed that a "retreating" contact angle ($\gamma_r$) (obtained by measuring while decreasing the drop volume by means of aspirating part of the drop's liquid with the syringe during, or immediately after the aspiration of liquid) measures the polar part of the solid surface. This is, unfortunately, pure nonsense: all one does while measuring $\gamma_r$, is observing the contact angle of a drop which is sessile on a surface that has just been contaminated by wetting with the contact angle liquid (van Oss, 1994, 2006). The erroneous impression of greater polarity, seemingly bequeathed by a lowering of the observed "retreating" contact angle, is simply due to the drop's liquid now resting on a surface that just has been wetted by that very same liquid. It is even possible to measure the degree of contamination of the solid surface through the undue wetting with the contact angle liquid, by using Cassie's equation for the interpretation of contact angle measurements on heterogeneous surfaces (consisting of a mosaic of two different materials) (Cassie and Baxter, 1944; Cassie, 1948):

$$\cos\theta_A = f_1 \cos\theta_1 + f_2 \cos\theta_2, \tag{2.16}$$

where $\theta_A$ is the average contact angle found on the heterogeneous surface, $f_1 + f_2 = 1$, and $\theta_1$ and $\theta_2$ are the contact angles pertaining to, respectively, pure component 1 and pure component 2 of the heterogeneous surface (van Oss, 1994, 2006).

Another widespread misapprehension is the belief that even when a finite contact angle with a drop of liquid, L, can be observed on a solid, flat, smooth surface, S, condensation of some of the vapor (emanating from the liquid, L) upon that solid surface forms a thin layer of liquid, L, around the drop, which then pushes at the drop from all sides, with a force named equilibrium spreading pressure ($\pi_e$). This effect would transform the Young–Dupré equation (Eq. (2.15)) into:

$$(1 + \cos\theta)\gamma_L = -\Delta G_{SL} - \pi_e. \tag{2.17}$$

This scenario was first proposed by Bangham and Razouk (1937) and has had many followers ever since. Now, when liquids do spread over a solid surface when $\gamma_L < \gamma_S$, there will exist a genuine, non-negligible spreading pressure, but no contact angle.

However, when $\gamma_L > \gamma_S$, which is the *conditio sine qua non* for the actual formation of finite (non-zero) contact angles, no significant spreading occurs (see also Fowkes et al., 1980). In 1998, van Oss, Giese and Wu demonstrated that the actual influence of deposited vapor-induced spreading pressures emanating from measurable drops with finite contact angles is exceedingly slight. Using surfaces of solid polyethylene oxide (PEO), which is very hydrophilic, as well as of polymethylmethacrylate (PMMA), which is moderately hydrophobic, it could be shown that the deviation caused by "$\pi_e$" in the observed contact angles ($\Delta\theta$) is less than 1°. It was only with drops of water that, under non-spreading conditions, the maximum deviation, $\Delta\theta$, on PMMA, was about 1.5° (van Oss, Giese and Wu, 1998), which is still of the same order of magnitude as the average experimental error inherent in contact angle measurement. These effects were also analyzed by using the Cassie equation (see above), to determine the percentage of the solid surface occupied by liquid deposited from the vapors emanating from the contact angle liquids' drops, which is the real cause of the (very slight) influence of condensed vapor emanating from the liquid of the drop. Again, using PMMA, it was found that depending on the contact angle liquid used, the percentage of the solid PMMA surface covered with condensate from the liquid drops, varied from 0.06 to 0.07% for glycerol and formamide, to 3.5% for water. The latter percentage for water could also routinely occur, depending on the ambient atmospheric humidity during the measurement.

It should also be stressed that such small effects as can be demonstrated to occur through drop vapor deposition onto solid surfaces are *not* due to layers of condensed liquid pushing at the drop (as schematically but erroneously illustrated by Bangham and Razouk, 1937), but are simply a consequence of an exceedingly modest degree of contamination (0.06 to 3.5%) of the solid surface by molecules emanating from the contact angle drop (van Oss et al., 1998). Thus, for all practical purposes, when real observable contact angles occur (i.e., when $\gamma_L > \gamma_S$), the dreaded effects of largely imaginary "equilibrium spreading pressures" caused by liquid pressures emanating from vapors coming from the liquid drops, raised by Bangham and Razouk (1937), may be safely ignored (see also Hauxwell and Ottewill, 1970; Good, 1975; Fowkes et al., 1980). This conclusion also makes the still persisting habit of using subscripts attached to $\gamma$, such as LV or SV (where "V" stands for "vapor") as superfluous as it is confusing.

Direct contact angle measurement, which is the easiest of all contact angle determination approaches, is best done by observation of the sessile drop through a small (10×) telescope, provided with a cross-hair in the rotating ocular lens holder which is calibrated into 360 degrees at its exterior periphery. The liquid is fed onto the precisely horizontal solid surface by means of a vertical syringe. The contact angle liquid has to be pure and, in most cases, should not be a solution. However, sub-molar aqueous concentrations of salt (e.g., 0.15 M NaCl, for use on biological materials) can be tolerated, as this salt content barely impinges on the surface tension of the liquid (van Oss et al., 1975). On the other hand, the presence of plurivalent ions, such as $Ca^{2+}$ may interact with certain surfaces; see Chapter 3, Section 6 and Chapter 8, Section 3, below.

It should also be noted that direct contact angle measurements may only be done on smooth solid surfaces. Rough surfaces tend to give rise to contact angles that are too high (see, e.g., Adamson, 1990). The principal reason for the undue increase in contact angle values measured on rough surfaces is that the contact angle liquid is only partly in contact with the solid material as it hovers for a significant portion of the rough surface over *air* that is interstitial between vertically protruding elevations. Now, *the water–air interface is the most hydrophobic surface known*, which is the main cause of the increased contact angle values observed on rough surfaces; see also Sub-section 1.2 in Chapter 11: The Water–Air Interface.

### 1.6.4 Contact angle measurement by thin layer wicking

Via wicking one can determine $\cos\theta$ on small particles. Wicking is the measurement of the speed of capillary advancement of liquid, L, through a cylindrical column, or along a flat surface coated with packed or contiguous solid particles. Crucial to the interpretation of the observed speed of capillary advancement of the liquid front is the Washburn (1921) equation (see also Adamson, 1990; Ku et al., 1985):

$$h^2 = (t \cdot R_a \cdot \gamma_L \cdot \cos\theta)/2\eta, \tag{2.18}$$

where h is the height (or distance) of capillary rise of liquid, L, in time, t; $R_a$ is the average radius of the air-filled interstices of pores and $\eta$ is the viscosity of liquid, L. This is essentially one equation with two unknowns: $R_a$ and $\cos\theta$. Thus, before being able to determine $\cos\theta$, $R_a$ must be measured. This is done by wicking with a low-$\gamma_L$ liquid, i.e., a *spreading liquid*, e.g., an alkane, such as octane, decane, dodecane, or tetradecane (see Table 2.1). As most inorganic particles and also many organic ones have a higher $\gamma_S$ than 29 mJ/m$^2$, all the aforementioned alkanes would spread on them. Also, as spreading liquids pre-wet solid surfaces (i.e., they form a "precursor film"; cf. de Gennes, 1990), $\cos\theta$ always exactly equals unity in these cases (van Oss, Giese, Li et al., 1992). In actual practice, the use of Washburn's equation (Eq. (2.18)) by counting the time t beginning at the very start of capillary advancement [i.e., from the precise moment the wicking tube (or plate) contacts the liquid] may be a source of some inaccuracy in the final results. This is because the act of bringing the capillary tube or the thin layer plate in contact with the liquid unavoidably causes some disturbance, so that it is not really possible to determine the point of t = 0 with precision. It is therefore preferable just to start measuring at an early but non-zero time, $t_1$ and to continue until a time, $t_2$, and use the time difference $(t_2 - t_1)$ which is easy to measure with precision and which obviates the influence of the disturbances inherent to the use of t = 0 as a starting point. One then uses the following version of the Washburn equation:

$$h_2^2 - h_1^2 = [(t_2 - t_1) \cdot R_a \cdot \gamma_L \cdot \cos\theta)]/2\eta. \tag{2.18A}$$

Meanwhile, wicking in glass capillaries filled with monosized spherical particles (see, e.g., Ku et al., 1985) yields excellent results. However with irregularly formed *polysized* clay or other mineral particles, wicking in vertical capillary tubes frequently results in an advancing front that is no longer impeccably horizontal, but quickly becomes skewed. This makes it virtually impossible to measure the precise height, h, of the front at any particular time, $t_n$. Following a suggestion by Professor Manoj K. Chaudhury, and based upon the technique of thin layer chromatography (using flat glass plates such as microscope slides, pre-coated with a contiguous layer of small particles), we started using such plates, coated with various types of clay or other mineral particles (van Oss, Giese, Li et al., 1992). With these thin layer wicking plates, used vertically, we could invariably observe advancing fronts that were straight and horizontal.

The main reason for the skewness of advancing liquid fronts obtained in capillaries filled with polysized irregular shaped mineral particles, lies in the fact that the rising liquid early on starts to lubricate the lower-lying particles, which tends to allow them to settle further down along a short distance, thus causing a partial air gap to develop between lower and higher particles, which invariably starts on one side of the capillary tube, and which then gives rise to a skewed liquid front which continues to advance while favoring one (vertical) side of the capillary tube's lumen. However, as thin layer wicking works with particles that sufficiently solidly adhere to the glass plate, no skewing develops when using that approach.

Direct contact angle measurement and wicking cannot normally be used interchangeably. For instance if one wishes to measure contact angles on clay particles, one is confronted with the fact that there are swelling clays (such as smectites, which swell in water and form very smooth surface layers upon air drying after having been deposited upon a flat plate from an aqueous suspension), so that due to their multi-directional swelling in various liquids swelling clays cannot be wicked, as the Washburn equation is only valid for the measurement of capillary liquid flow in one direction. Non-swelling clays on the other hand, which upon drying continue to present a *rough surface* cannot be used for direct contact angle determination. Thus swelling clays cannot be wicked for the purpose of obtaining their $\cos \theta$ value and for non-swelling ("rough") clay particles, the value of their contact angle cannot be determined by direct measurement of a sessile drop. This dilemma, which lasted approximately from 1991 to 1994, made it impossible to compare the two different measurement methods for the purpose of verifying whether both approaches would indeed allow one to arrive at the same numerical value for $\theta$, when applied to the same material. However, as published in 1995 by Costanzo et al., a method was developed by which a comparison could be made between direct contact angle measurement and the determination of $\cos\theta$ by (thin layer) wicking, using synthetic monosized cuboid hematite particles in both approaches. These synthetically formed hematite particles are monosized *cubes* with sides of approximately 600 nm. When deposited on a glass slide by allowing the particles to sediment onto it from an aqueous suspension, followed by air-drying, an extremely smooth, self-organized flat, shiny monolayer of hematite particles was obtained which could serve as a rigorously flat layer for direct contact angle measurements, as well as for the determination of $\cos\theta$ via thin layer wicking. Within experimental error, the contact angles, as well as the $\gamma^{LW}$, $\gamma^{+}$ and $\gamma^{-}$ values obtained for these hematite layers were closely comparable; see Table 2.6.

Thus the new methodology of thin layer wicking (Giese et al., 1996; Costanzo et al., 1991; van Oss, Giese, Li et al., 1992) made it possible for the first time to determine the surface-thermodynamic properties of dozens of clay and other mineral particles (Giese et al., 1996; Giese and van Oss, 2002).

## 2. The Values for $\gamma^{LW}$, $\gamma^{+}$ and $\gamma^{-}$ for Water at 20 °C

When, during the second half of the 1980's, we worked on developing the combining rules for the surface-thermodynamic interactions among polar molecules, particles and surfaces while immersed in a polar liquid such as water, which combining rules had to be based on Lewis acid–base (or electron-acceptor/electron-donor) interactions, we looked in vain for a stable compound or material that could be used as a reference standard. Even in Gutmann's work (1978) the electron-acceptor and electron-donor "numbers" given in Gutmann's Tables 2.1 and 2.3 gave no convincing acceptor or donor "numbers" that would allow one to identify standard values for polar materials that could be adopted as quantitative standards, expressed in S.I. units (see, e.g., van Oss, 1994, note on pp. 27, 28; 2006, note on p. 34). It therefore became necessary to establish such a standard. To that effect water, at 20 °C, appeared to be the obvious choice to serve as the standard polar reference material. It had already been known since 1963 (Fowkes, 1963, 1964, 1965) that with water at 20 °C its value for $2\sqrt{(\gamma^{+}\cdot\gamma^{-})}$ equals 51 mJ/m$^2$ (see Eq. (2.3), above). It was also obvious from Gutmann (1978) that for liquid water both its electron-accepticity and its electron-donicity had sizable values and could be estimated not to be very different from one another. This is in contrast with non-aqueous polar hydrophilic materials, which frequently turned out to be monopolar electron-donors (van Oss, Chaudhury and Good, 1987). Furthermore, given that liquid water appeared to have rather comparable electron-acceptor and electron-donor potentials (see above) and that their *product*, $\gamma^{+} \times \gamma^{-}$, is already known for water, it finally remained only necessary to establish their *ratio*, $r = \gamma^{+}/\gamma^{-}$. It was then decided to assume this ratio for water to be unity at 20 °C (van Oss, Chaudhury and Good, 1987, 1988), i.e.: $r = \gamma^{+}/\gamma^{-} = 1.0$. Thus, the polar Lewis acid–base reference standard was designated to be liquid water, at 20 °C and was established as:

$$\gamma_W^{+} = \gamma_W^{-} = 25.5 \text{ mJ/m}^2. \tag{2.19}$$

Via subsequent measurements the $\gamma^{+}$ and $\gamma^{-}$ values for other polar condensed-phase materials were expressed with reference to the above standards for water (for many examples, see Section 3, below). It is easily shown (van Oss, Chaudhury and Good, 1987, 1988; van Oss, 1994, 2006) that whilst the $\gamma^{+}$ and $\gamma^{-}$ values of other polar materials are all expressed as relative to the $r = 1.0$ definition for water at 20 °C (and would be different if one were to assume a different value for r), as a consequence of the fact that, e.g., $\gamma_x^{+}$ and $\gamma_y^{-}$, are everywhere only used in terms expressed as $\sqrt{(\gamma_x^{+}\cdot\gamma_y^{-})}$, the values for $\gamma_1^{AB}$, $\gamma_{12}^{AB}$, $\Delta G_{12}^{AB}$, $\Delta G_{11}^{AB}$, $\Delta G_{121}^{AB}$ and $\Delta G_{132}^{AB}$ are nonetheless the *absolute values*, as obtained via the Dupré equations ((2.9)–(2.11) and (2.13)), given above.

It would also have been possible to obviate the dependence of the values of $\gamma_x^{+}$ and $\gamma_x^{-}$ found for various polar materials, x, on the $r = 1.0$ assumption for $\gamma_W^{+}/\gamma_W^{-}$ for water at 20 °C by using dimensionless entities, i.e., $\delta_{1W}^{+}$ and $\delta_{1W}^{-}$, where:

$$\delta_{1W}^{+} = \sqrt{(\gamma_1^{+}\cdot\gamma_W^{+})} \tag{2.20A}$$

and:

$$\delta_{1W}^{-} = \sqrt{(\gamma_1^{-} \cdot \gamma_W^{-})} \tag{2.20B}$$

(van Oss, Chaudhury and Good, 1987; 1988; van Oss, 1994, 2006). However, in consideration of the fact that this approach would be much less familiar and user-friendly than the simple $\gamma^+$ and $\gamma^-$ method explained earlier in this chapter, we felt that it would be counterproductive to insist on a continued use of the dimensionless but more complicated and unfamiliar $\delta_{1W}^{+}$ and $\delta_{1W}^{-}$ approach.

It is important to note that $r = \gamma^+/\gamma^- = 1.0$ is only valid when the temperature of water is 20 °C. At all other water temperatures that ratio is different: at lower temperatures $r < 1.0$ and at temperatures higher than 20 °C, $r > 1.0$. For instance, at 38 °C, $r = \gamma_W^{+}/\gamma_W^{-}$ was found to be equal to 1.75 (van Oss, 1994, p. 301; 2006, pp. 95–96). The influence of temperature on the value of r for water is further discussed in Chapter 5, (Sub-section 1.4) and in Chapter 9.

## 3. Apolar and Polar Surface Properties of Various Other Condensed-Phase Materials

In the following six tables the values are given for $\gamma_S$ (or $\gamma_L$), $\gamma_{LW}$, $\gamma^+$, $\gamma^-$, as well as for $\Delta G_{SWS}$ or $\Delta G_{LWL}$, determined at 20 °C, and expressed in mJ/m$^2$ units. They are all given relative to the standard ratio, $r = \gamma^+/\gamma^- = 1.0$ for water at 20 °C.

### 3.1 Liquids

Table 2.1 shows the $\gamma$ values as well as $\Delta G_{LWL}$ for a number of apolar and polar liquids. Among the listed alkanes the first three are mainly used (in the context of this chapter) as spreading liquids needed for the determination of $R_a$ [i.e., the average interstitial pore radius which is used in wicking; cf. the Washburn equation (Eq. (2.18))]. The viscosities of these liquids at 20 °C can be found in, e.g., the CRC Handbook of Chemistry and Physics, or in van Oss (2006, Chapter 17). The $\gamma$ values of the liquids used in direct contact angle determinations, as well as in wicking (diiodomethane, water, glycerol, formamide and ethylene glycol) are also found in Table 2.1 and their viscosities can be found as cited above. Furthermore, the $\gamma$ values for hexadecane, chloroform, ethyl ether and ethyl acetate are given: these are used in microbial adhesion to solvents (MATS) measurements (Bellon-Fontaine et al., 1996). Hexadecane is also used in microbial adhesion to hydrocarbons (MATH); see Rosenberg et al. (1980); for MATS and MATH, see also Chapter 6, Sub-section 2.1.2.

Finally, the polar $\gamma$ values as well as the $\Delta G_{LWL}$ values of diiodomethane, chloroform, benzene, ethyl ether and ethyl acetate have been newly determined from their solubility data; see van Oss, Docoslis, Giese (2001, 2002); van Oss, Giese, Docoslis (2001); van Oss, Wu et al. (2001); van Oss, Giese, Good (2002).

**Table 2.2** $\gamma$, as well as $\Delta G_{SWS}$ values for synthetic polymers, at 20 °C, in mJ/m²

| Polymer | $\gamma_S$ | $\gamma^{LW}$ | $\gamma^+$ | $\gamma^-$ | $\gamma^{AB}$ | $\Delta G_{SWS}$ |
|---|---|---|---|---|---|---|
| Teflon[a] | 17.9 | 17.9 | 0 | 0 | 0 | −102.4 |
| Polyisobutylene[b] | 25.0 | 25.0 | 0 | 0 | 0 | −102.2 |
| Polyethylene[c] | 33.1 | 33.0 | 0 | 0 | 0 | −104.4 |
| Polypropylene[b] | 25.7 | 25.7 | 0 | 0 | 0 | −102.4 |
| Nylon 6.6[d] | 37.7 | 36.4 | 0.02 | 21.6 | 1.3 | −11.6 |
| Polymethylmethacrylate[b] | 40.6 | 40.6 | 0 | 12.0 | 0 | −37.7 |
| Polystyrene[b] | 42.0 | 42.0 | 0 | 1.1 | 0 | −87.4 |
| Cell culture quality polystyrene[c] | 46.8 | 43.7 | 0.12 | 20.0 | 3.1 | −18.4 |
| Polyvinyl chloride[b] | 43.8 | 43.0 | 0.04 | 3.5 | 0.75 | −68.8 |
| Polyvinyl pyrrolidone[c] | 43.4 | 43.4 | 0 | 29.7 | 0 | +0.72 |
| Polyethylene oxide (6000)[e] | 43.0 | 43.0 | 0 | 64.0 | 0 | +52.5 |

*a* Data from Chaudhury (1984).

*b* Data from van Oss, Chaudhury, Good (1989); van Oss, Ju et al. (1989).

*c* Data from van Oss (1994, 2006).

*d* Data from van Oss, Good and Busscher (1990).

*e* Data from van Oss (1994, 2006).

## 3.2 Synthetic polymers

The $\gamma$ values for a number of frequently used solid polymer surfaces are given in Table 2.2. Their degrees of hydrophobicity or hydrophilicity are expressed as $\Delta G_{SWS}$, where the degree of hydrophobicity is given by the values of $\Delta G_{SWS} < 0$ and *idem* of hydrophilicity by $\Delta G_{SWS} > 0$ (van Oss and Giese, 1995; see also Chapter 5, Sections 1.1 and 1.2). With the exception of the two last polymers listed in Table 2.2 (PVP and PEO) all the other polymers are insoluble in water (cf. their strongly negative $\Delta G_{SWS}$ values). PVP and PEO on the other hand have strongly positive $\Delta G_{SWS}$ values and thus are very soluble in water; see also Chapter 7, Section 3.

## 3.3 Plasma proteins (see Table 2.3)

Fibrin (which is not a plasma protein *sensu stricto*, but the ultimate end-product of the plasma protein, fibrinogen, after clotting) is an indispensable factor in blood coagulation [which is why it is helpful for fibrin to be hydrophobic (van Oss, 1990)], hence the negative value of its $\Delta G_{SWS}$.

In their normal hydrated state the major mammalian blood plasma proteins that are shown here, i.e., HSA, IgG and fibrinogen) are quite soluble in water, as exemplified by their positive $\Delta G_{SWS}$ values at neutral pH. However, once air-dried, HSA as well as IgG become hydrophobic (i.e., $\Delta G_{SWS} < 0$). This is primarily due to the influence of the water–air interface, which is completely

**Table 2.3** $\gamma$, as well as $\Delta G_{SWS}$ values for plasma proteins, at 20 °C, in mJ/m²

| Plasma protein | $\gamma_S$ | $\gamma^{LW}$ | $\gamma^+$ | $\gamma^-$ | $\gamma^{AB}$ | $\Delta G_{SWS}$[e] |
|---|---|---|---|---|---|---|
| Human serum albumin (HSA) dry, pH 4.8[a] | 45.0 | 44.0 | 0.03 | 7.6 | 0.95 | −26.2 |
| HSA, dry, pH 7[a] | 41.4 | 41.0 | 0.002 | 20.0 | 0.4 | −17.5 |
| HSA, hydrated, 1 layer of water of hydration, pH 7[b] | 27.6 | 26.6 | 0.003 | 87.5 | 1.03 | +85.5 |
| Human IgG, dry, pH 7[a] | 45.2 | 42.0 | 0.3 | 8.7 | 3.2 | −44.4 |
| IgG, hydrated, pH 7[a] | 51.3 | 34.0 | 1.5 | 49.6 | 17.3 | +26.1 |
| Bovine fibrinogen, dry[c] | 40.3 | 40.3 | 0 | 53.2 | 0 | +39.7 |
| Bovine fibrin, dry[d] | 44.0 | 40.2 | 0.3 | 12.0 | 3.8 | −34.1 |

*a* Data from van Oss (1989a, 1989b).

*b* Data from van Oss and Goog (1988).

*c* Data from van Oss (1994, 2006).

*d* Data from van Oss (1990, 1991a, 1991b).

*e* Obtained by using Eq. (2.11), for $\Delta G_{SWS}$.

hydrophobic (see Chapter 11, Sub-section 1.3). Thus, upon air-drying, the hydrophobic interiors of the exteriorly hydrophilic globular protein molecules such as HSA and IgG turn outward to the air-side and stay there until dry. This is evident from the negative $\Delta G_{SWS}$ values of the surfaces of air-dried HSA and IgG, even though in their normal, hydrated state, HSA and IgG are quite hydrophilic. It can also be seen in Table 2.3 that close to its isoelectric pH of 4.8, dried HSA is even more hydrophobic than dried HSA at pH 7. In contrast with the globular proteins, HSA and IgG, the (bovine) plasma protein, *fibrinogen* does not appear to be influenced by air-drying insofar that this process does not make it hydrophobic. Human fibrinogen (not shown in Table 2.3) also does not significantly change its degree of hydrophilicity upon air-drying (van Oss, 1994, 2006). It should however be realized that fibrinogen is not a globular protein but has, instead, a long, thin cylindrical structure, without a hydrophobic interior.

In Table 2.3 on also clearly sees the influence of hydration on the surface properties of the serum proteins, HSA and IgG. When dry these proteins have a $\gamma^{LW}$ value in the 40's (mJ/m²) whereas when hydrated these $\gamma^{LW}$ values are much decreased (to 26.6 and 34.0 mJ/m², respectively), in the direction of $\gamma_w^{LW} = 21.8$ mJ/m² for water. When dry both serum proteins are clearly hydrophobic, with a $\gamma^-$ well below 28 mJ/m² and with a negative $\Delta G_{SWS}$ value. Both proteins are clearly completely or almost completely $\gamma^-$ monopoles, with a significant $\gamma^-$ and with a $\gamma^+$ value which is close to zero, or at least quite low, even when hydrated. This shows that the first layer of water molecules of hydration of these proteins is practically entirely bound via their electron-acceptors to the electron-donating part of the proteins' surfaces.

**Table 2.4** $\gamma$, as well as $\Delta G_{SWS}$ values for sucrose, glucose and dextran, in the dry as well as in the dissolved state in water (extrapolated to 100% carbohydrate) at 20 °C in mJ/m$^2$

| Carbohydrates | | $\gamma_S$ | $\gamma^{LW}$ | $\gamma^+$ | $\gamma^-$ | $\gamma^{AB}$ | $\Delta G_{SWS}$ |
|---|---|---|---|---|---|---|---|
| Sucrose | Dry[a] | 41.6 | 41.6 | 0 | 59.5 | 0 | +47.5 |
| Glucose | Dry[a] | 42.2 | 42.2 | 0 | 51.1 | 0 | +35.7 |
| Dextran | Dry[a] | 42.2 | 42.2 | 0 | 55.0 | 0 | +41.2 |
| Sucrose | In aqueous solution[b] | 141.8 | 41.6 | 28.5 | 88.0 | 100.2 | −11.4 |
| Glucose | In aqueous solution[b] | 150.9 | 42.2 | 34.5 | 85.6 | 108.7 | −20.5 |
| Dextran | In aqueous solution[b] | 42.0 | 63.4 | 2.0 | 57.0 | 21.4 | +29.8 |

a Data from van Oss (1994, 2006).

b Data from Docoslis et al. (2000; extrapolated to 100% carbohydrate).

## 3.4 Carbohydrates

In Table 2.4 two normal, low molecular weight sugars: sucrose and glucose are shown, as well as dextran T-150, which is a polymer of maltose, which in turn is a dimer of glucose. The surface properties of these carbohydrates are shown in the upper part of the table as the dried di- and mono-saccharides and as the dried polysaccharide, dextran. It is clear that the surfaces of all three are *monopolar* electron-donors (i.e., their $\gamma^-$ is large and their $\gamma^+$ is zero). In the *dried state* the simple sugars as well as the polysaccharide are monopolar electron donors. In *aqueous solution*, however, both the simple sugars and the polysaccharide are *dipolar*. With dextran this dipolarity is not very pronounced, but with the two simple sugars the dipolarity in the dissolved state is extremely significant, to the point where concentrated sugar solutions in water strongly enhance the polar free energy of cohesion of water and thus also its hydrophobizing capacity; see Chapter 5, Subsection 1.5. The glucose polymer, dextran, does not display the strong polar free energy of cohesion of its monomer, glucose, because in the process of covalent polymerization the dextran chains have lost the greater part of the polar free energy of their erstwhile single glucose molecules.

## 3.5 Clays and other minerals

The mineral surfaces or particles shown in Table 2.5 are all metal oxides which is most likely why their $\gamma^{LW}$ values only vary between about 31 and 44 mJ/m$^2$. Their polar ($\gamma^{AB}$) values on the other hand vary tremendously, i.e., their electron-donicities ($\gamma^-$) (these mineral particles are all close to being monopolar) vary between 3.6 and 51.5 mJ/m$^2$. With monopolar electron-donating surfaces, $\gamma^-$ is a fairly good indicator of hydrophilicity, although that property is more rigorously defined by a positive sign and an elevated value for $\Delta G_{SWS}$. In Table 2.5 the smectite, bentolite-L, mica and $SiO_2$ are the only hydrophilic samples, which for monopolar electron-donors with a $\gamma^{LW}$ of roughly 40 mJ/m$^2$ correlates with a $\gamma^-$ that is greater than about 28.5 mJ/m$^2$, which is clearly the case with the three minerals mentioned

**Table 2.5** $\gamma$, as well as $\Delta G_{SWS}$ values for various clays and related minerals, at 20 °C, in mJ/m²[a]

| Mineral | Contact angle measurement method | $\gamma_S$ | $\gamma^{LW}$ | $\gamma^+$ | $\gamma^-$ | $\gamma^{AB}$ | $\Delta G_{SWS}$[d] |
|---|---|---|---|---|---|---|---|
| *Swelling clays (smectites)* | | | | | | | |
| Bentolite-L (commercial smectite) | DCA[b] | 57.6 | 44.1 | 1.0 | 45.3 | 13.5 | +19.5 |
| SWY-1 (So. Wyoming) | DCA | 53.9 | 40.7 | 1.5 | 29.2 | 13.2 | −0.44 |
| Hectorite | DCA | 39.9 | 39.9 | 0 | 23.7 | 0 | −9.1 |
| *Non-swelling clays* | | | | | | | |
| Kaolinite | TW[c] | 47.3 | 39.9 | 0.4 | 34.3 | 7.4 | +8.8 |
| Talc particles (Fisher) | TW | 37.2 | 30.7 | 1.8 | 5.9 | 6.5 | −40.4 |
| Pyrrophillite | TW | 39.7 | 33.9 | 1.7 | 4.9 | 5.8 | −45.2 |
| *Other metal oxides* | | | | | | | |
| Mica (muscovite) | DCA | 59.8 | 40.6 | 1.8 | 51.5 | 19.2 | +24.9 |
| $SiO_2$ particles (Fisher) | TW | 50.7 | 39.2 | 0.8 | 41.4 | 11.5 | +17.9 |
| $ZrO_2$ particles | TW | 39.1 | 34.8 | 1.3 | 3.6 | 4.3 | −50.2 |

*a* Selected data from Giese et al. (1996); see also Giese and van Oss (2002).

*b* DCA = direct contact angle measurements.

*c* TW = tin layer wicking measurements.

*d* Calculated using Eq. (2.11).

above. The smectite, SWy-1 (with a $\gamma^-$ of 29.2 mJ/m²) is a limiting case: it is really extremely close to being hydrophilic, with a $\Delta G_{SWS}$ of just −0.44 mJ/m². [It should be noted that SWy-1 also has a (small) $\gamma^+$ value, of 1.5 mJ/m²: this increases the value of $\gamma^-$ beyond which the material would be hydrophilic]. What contributes to bringing SWy-1 even closer to behave as a hydrophilic entity is due to the fact that it also has a sizeable $\zeta$-potential, of −60.1 mV (Giese et al., 1996), which helps it in allowing its particles to form stable suspensions in water; see also Chapters 3 and 8.

It should also be noted that among all these clay and other mineral surfaces and particles there is a wide variability in hydrophilicity/hydrophobicity (cf. the $\Delta G_{SWS}$ column of Table 2.5), which is true for the swelling as well as for the non-swelling clays. Among the other metal oxides, mica is very hydrophilic, as is $SiO_2$ (silica) whilst $ZrO_2$ (zirconia) is extremely hydrophobic. With these oxides of tetravalent metals, for the lower Mw cations (e.g., $Si^{4+}$), hydrophilicity obtains, whilst zirconia with the high Mw $Zr^{4+}$ is decidedly hydrophobic, as is also $SnO_2$ (see Giese et al., 1996, and also van Oss and Giese, 1995; Giese and van Oss, 2002). $TiO_2$ is also hydrophilic according to Giese et al. (1996), but depending on how Titania is formed, it can also assume a somewhat more hydrophobic form.

**Table 2.6** (A) Comparison between the $\gamma$ as well as the $\Delta G_{SWS}$ properties of ground solids (using thin layer wicking) and these properties measured on the original solid surfaces (using direct contact angle measurements), as contrasted with (B); The same comparison, using identical monosized synthetic cuboid hematite particles (all values in mJ/m$^2$, at 20 °C)

| Solids | Method | $\gamma_S$ | $\gamma^{LW}$ | $\gamma^+$ | $\gamma^-$ | $\gamma^{AB}$ | $\Delta G_{SWS}$ |
|---|---|---|---|---|---|---|---|
| *(A) Large solids vs ground solids*[a] | | | | | | | |
| Dolomite | DCA[c] | 42.5 | 37.6 | 0.2 | 30.5 | 4.9 | +4.5 |
| Ground dolomite | TW[d] | 20.4 | 27.1 | 0.2 | 13.6 | 3.3 | −12.5 |
| Glass | DCA | 51.7 | 33.7 | 1.3 | 62.2 | 18.0 | +41.8 |
| Ground glass | TW | 38.8 | 31.1 | 0.4 | 37.1 | 7.7 | +16.8 |
| *(B) Cubic hematite particles*[b] | | | | | | | |
| Hematite particles | DCA | 53.4 | 45.6 | 0.3 | 50.4 | 7.8 | +28.2 |
| Hematite particles | TW | 50.6 | 46.1 | 0.1 | 50.1 | 4.5 | +29.4 |

*a* Data from Giese et al. (1996); see also Wu et al. (1996); Giese and van Oss (2002).

*b* Data from Costanzo et al. (1995).

*c* DCA: Direct contact angle measurements.

*d* TW: Contact angles obtained via thin layer wicking (see, e.g., van Oss, Giese, Li et al. 1992).

## 3.6 Large solid surfaces vs ground solids—Direct contact angle measurements vs thin layer wicking

### 3.6.1 Large solid surfaces vs ground solids

Table 2.6(A) shows that the electron-donicity of hydrophilic minerals is strongly decreased upon grinding [these are only two examples out of a larger number of cases; see Giese et al. (1996); Wu et al. (1996)]. Now, the experimental methodology requires that for flat solid surfaces direct contact angle measurements be utilized for the determination of their surface-thermodynamic values. However with ground particles that approach is not recommended because of the roughness of layers of small solid particles deposited upon a flat surface; see Sub-section 1.6, above. Thus, in the latter case, thin layer wicking is the preferred (and essentially the only) methodology. We therefore wished to be reassured that the apparent differences in $\gamma$ properties found after grinding mineral solids, could not have been caused by the different measurement approaches used in determining contact angles on large, smooth, solid surfaces vs on ground particles. To that effect mineral particles were synthesized which could be used *for direct contact angle determinations* on smooth flat surfaces, *as well as* for the determination of $\cos\theta$ on the same surfaces *by thin layer wicking*; see the next sub-section.

### 3.6.2 Comparison between direct contact angle measurement and contact angle determination by thin layer wicking

Monosized cuboid hematite particles of 0.6 μm were synthesized. These could be deposited from an aqueous suspension onto glass slides and dried, to form an exceedingly smooth shiny monolayer of adjoining hematite cubes. With these thin layers of hematite direct contact angle determinations (DCA) as well as thin layer wicking (TW) could be effected (Costanzo et al., 1995). Table 2.6(B) shows that the results were closely comparable; see also Sub-section 1.6.4, above.

### 3.6.3 Conclusions *re* the grinding of metal oxide solids

Given the results obtained with layers of synthetic cuboid hematite particles (Costanzo et al., 1995; see the preceding sub-section, above) it may be concluded that the direct contact angle and the wicking method both measure contact angles accurately so that we may state that *upon grinding*, metal oxide solids become much less hydrophilic, or even change from hydrophilic to hydrophobic (see Table 2.6(A)). The grinding was done slowly in air, with an ordinary mortar and pestle which was mechanically driven but caused no significant heating effects (Wu et al., 1996). Various other metal oxides, not shown in these tables, also become more hydrophobic when ground, or when ground more finely, e.g.: talc, calcite, silica, zirconia (Wu et al., 1996).

The likely mechanism of this type of hydrophobization through grinding would be as follows: In Nature there is an excess of predominantly electron-donating surfaces (van Oss et al., 1997) in contrast with a prevailing scarcity and often a virtual absence of electron-accepticity on all surfaces measured so far. Nonetheless, inside the bulk of solid metal oxide materials much of the non-covalent part of the free energy of cohesion appears to comprise electron-acceptor/electron-donor complexes, whilst the excess of unbound or more weakly bound electron-donors move toward the solid–air interface. When present at that interface at fairly high concentrations, the surfaces of such materials are experimentally determined to be hydrophilic and when the electron-donicity of such surfaces is less concentrated, they are measured as being hydrophobic. Now, when a solid piece of an apparently hydrophilic metal oxide is ground into small particles, its total surface area is vastly increased, which causes the excess electron-donors to become distributed over a larger surface area, which results in a decrease in the surface density of eletron-donors. This in turn gives rise to a less hydrophilic surface (see the ground glass in Table 2.6(A)), or even to a frankly hydrophobic surface, as is the case with ground dolomite (Table 2.6(A)).

CHAPTER THREE

# The Extended DLVO Theory

## Contents



*Interface Science and Technology*, Volume 16
ISSN 1573-4285, doi: 10.1016/S1573-4285(08)00203-2

# 1. Hamaker Constants and the Minimum Equilibrium Distance between Two Non-Covalently Interacting Surfaces of Condensed-Phase Materials

## 1.1 Hamaker constants and their relation to $\gamma^{LW}$ and the minimum equilibrium distance, $d_0$, between two surfaces of condensed-phase materials

In 1968 Fowkes had already noted the connection between the surface tensions of non-polar molecules and their van der Waals–London (Hamaker) constants; see also Hough and White (1980). This connection involves the minimum equilibrium distance ($d_0$) between two such molecules, *in vacuo*, which Israelachvili first used in 1973, to arrive at a value for $d_0$ of about 0.2 nm; see also Israelachvili (1974). In his book of 1985, Israelachvili expressed this connection as:

$$\gamma = A/24\pi d_0^2 \tag{3.1}$$

for apolar systems. Here, $\gamma$ is the surface tension of the apolar compound in question and A is the van der Waals, or Hamaker constant. During the 1980's the calculation of Hamaker constants had become somewhat more accurate for apolar systems and by that time the Lifshitz theory was increasingly taken into account. Thus, Israelachvili (1985) found a value for $d_0$ of 0.165 nm, using the Lifshitz approach (Lifshitz, 1955). However, at least up to 1972, when Visser gave lists of published Hamaker constants from many different sources, these often varied greatly for the same compounds, dependent on how they had been derived. By the late 1980's it had become easier to determine the Lifshitz–van der Waals component of the surface tension ($\gamma^{LW}$) of condensed-phase materials (see also van Oss, Chaudhury and Good, 1987, 1988), so that one could redefine Eq. (3.1) more accurately as:

$$\gamma^{LW} = A/24\pi d_0{}^2. \tag{3.2}$$

This allowed van Oss, Chaudhury and Good (1988), while comparing Hamaker constants derived from spectroscopic constants and permittivities with those obtained from $\gamma^{LW}$ values, to arrive at a value for $d_0 = 0.157$ nm [±0.009 nm (SD)], averaging the results obtained with a variety of compounds, including five gases, Teflon, polystyrene, polymethylmethacrylate, six alkanes, benzene, chlorine, $CS_2$, glycerol and mercury (see also van Oss (1994, 2006).[1] The presence of a small but finite distance between all condensed-phase materials so that they can only

[1] The surface tension properties of glucose (dissolved in water) are: $\gamma^{LW} = 42.2$, $\gamma^+ = 34.5$ and $\gamma^- = 85.6$; all in mJ/m$^2$ (Docoslis et al., 2000). See Eq. (2.13) for obtaining $\Delta G_{132}$, or in this case, $\Delta G_{swa}$ (where subscript "a" stands for air).

approach each other to a given minimum equilibrium distance (with a length, $d_0 = 0.157$ nm) is caused by the Born repulsion, which is exceedingly strong at very short distances and does not allow a closer approach than $d = d_0$.

It should further be noted that Israelachvili (1985, in his book p. 158) showed Hamaker constants which were not only based on the $\gamma$ values of (non-polar) alkanes but also in a number of cases on the total $\gamma$ values of polar compounds. Nonetheless his resulting $d_0$ values appear remarkably close to those obtainable when using the $\gamma^{LW}$ instead of the total $\gamma$ values of polar compounds (see Eq. (3.2)). The reason for this probably lies in the fact that the value of $d_0$ not only applies to LW forces but also to the two other non-covalent interaction types, i.e., to Lewis acid-base (AB) and electrical double layer (EL) forces as well and it also should be noted that the use of $d_0$ remains compatible with calculations based on Hamaker's equation for all macroscopic non-covalent systems (see Hamaker, 1937a, Eq. (20)) which also describes macroscopic-scale interaction energies between two semi-infinite plane parallel plates at "contact":

$$\Delta G^{LW} = -A/12\pi d_0^{\,2} \qquad (3.3)$$

(see Sub-section 2, below). It should be recalled here that $\Delta G^{LW} = -2\gamma^{LW}$; cf. Eq. (2.4).

## 1.2 The minimum equilibrium distance, $d_0$, as a constant

In the preceding sub-section it has been shown that the minimum equilibrium distance ($d_0$) between two condensed-phase surfaces which attract one another via Lifshitz–van der Waals forces equals approximately 0.157 nm [(±0.009 nm (SD)]. Now, the fact that the same $d_0$ value which was found to obtain between non-polar entities was also found via the Lifshitz–van der Waals part of the attraction between polar surfaces, as well as between metallic surfaces (e.g., mercury), would indicate that the distance, $d_0$ is not altered by the presence of polar (AB), nor by metallic interaction energies (or, by extension, by EL energies). One may therefore conclude that the $d_0$ found for attractive LW energies remains essentially the same as the minimum equilibrium distance to which attractive non-covalent polar (Lewis acid–base or AB) energies and/or attractive metallic and EL energies would also allow two condensed-phase surfaces to approach one another. It therefore appears reasonable to propose that $d_0 \approx 0.157$ nm may be used as a *general constant*, representing the minimum equilibrium distance between two condensed-phase surfaces when attracted to one another by any one of the three non-covalent interactions, i.e., by Lifshitz–van der Waals, Lewis acid-base and electrical double layer forces.

At surface-to-surface distances of the order of ≈5 nm, the minimum equilibrium distance, $d_0 \approx 0.157$ nm plays a non-negligible role, so that it may be more accurate to designate the surface-to-surface distance as ($d$–$d_0$) for all distances, $d < 5$ nm. For distances greater than about 5 nm this measure is of lesser impact.

Especially with hydrophilic, hydrated proteins in aqueous solution [e.g., human serum albumin (HSA)] with one layer of tightly bound water of hydration (van Oss, 1994, 2006) there is a $\Delta G_{SWS}$ repulsion in water, which causes the protein molecules to remain dissolved by keeping them apart, so that the minimum equilibrium distance, $d_0$, is normally not attainable. The closest two such hydrated protein molecules can approach one another will then be to within one $d_0$ thickness on each side, plus the thickness of one layer of water molecules ($\approx$0.3 nm) from the liquid medium, i.e., $0.157 + 0.3 + 0.157 = 0.61$ nm. This is not the distance between the naked surfaces of, e.g., two HSA molecules, but between two tightly *hydrated* HSA molecules, in other words, between their respective layers of strongly bound water of hydration (one layer of tightly bound water of hydration per HSA molecule; see van Oss, 1994, 2006). This analysis supersedes a more speculative earlier treatment of the nature of $d_0$, by van Oss and Good (1984).

### 1.3 The proportionality constant $A_{ii}/\gamma_i^{LW}$

In general the Hamaker constant for a given material, i, that is, $A_{ii}$ (see Eq. (2.2)), is most easily determined via contact angle measurements on material, i (see Chapter 2, Sub-section 1.6), to allow one to obtain the value for $\gamma_i$. As the next step a simple proportionality constant linking A and $\gamma^{LW}$ can be used, of which constant the minimum equilibrium distance, $d_0$, is a component. Taking $d_0 \approx 0.157$ nm, the proportionality constant (see Eq. (2.2)) is found as follows:

$$A_{ii} = 1.8585\ (\pm 0.0065\ \text{SD}) \times 10^{-21} \gamma_i^{LW}, \tag{3.4}$$

where $A_{ii}$ (i.e., the Hamaker constant for the Lifshitz–van der Waals interaction between two semi-infinite plane parallel surfaces of material i, *in vacuo*, expressed in J) and $\gamma_i^{LW}$ is the Lifshitz–van der Waals component of material, i, expressed in $mJ/m^2$ (van Oss, 1994, 2006).

However, as a consequence of a lack of detailed spectroscopic data, causing a degree of uncertainty inherent in the Lifshitz calculations, the overall variability of the proportionality constant ($A_{ii}/\gamma_i^{LW} = 1.8585 \times 10^{-21}$), when used to determine the Hamaker constant from a measured $\gamma_i^{LW}$ value, has been estimated at $\pm$8% (van Oss, Chaudhury and Good, 1988), which is comparable to the variability of $A_{ii}$ values obtained via the Lifshitz approach when using spectroscopic data, see also Hough and White (1980).

## 2. The DLVO Theory Extended by the Addition of Polar Interaction Energies Occurring in Water

### 2.1 DLVO and XDLVO theories

If one looks at an original version of what is now called the DLVO theory, e.g., at Verwey and Overbeek (1948), or to a precursor of that theory (Hamaker, 1936, 1937a, 1937b), one notes that the treatment of electrical double layer interactions concentrates on interactions taking place in a liquid which is not actually called water, but which has properties that are reminiscent of those of water, especially in the treatment of electrical double layer interactions. In the classical DLVO theory electrical double layer *repulsions* represented one half of the interactions studied and these repulsions closely resembled those occurring in aqueous systems. The counterpart of these electrostatic repulsions treated in the classical DLVO theory were those of van der Waals–London *attractions* and the interplay between these two remained, for over half a century and in some publications up to the present, the only two forces under consideration.

Neither Hamaker (1936) nor Verwey and Overbeek (1948) clearly identified the liquid they had in mind, while designating the colloids immersed in it as "lyophobic." However Hamaker (1936) at least hinted at water by postulating that with lyophobic colloids the influence of hydration may be neglected. [Which, incidentally, it may not: even the most "lyophobic" (e.g., hydrophobic) surfaces will become hydrated when immersed in water to the extent of: $\Delta G_{\text{hydration}} \approx -40\ \text{mJ/m}^2$, which is not negligible at all]. Also, "lyophobic" (or, when in water, "hydrophobic") particles, when immersed in a polar, even mildly self-hydrogen-bonding liquid will be flocculated through a Lewis acid–base driven "solvophobic" (or hydrophobic) effect. Furthermore, if one uses really non-polar liquids, one would be unable to dissolve sufficient amounts of neutral salts in them to need to worry greatly about the influence of ionic strengths on the Debye thickness of the Gouy–Chapman diffuse double layer. Finally, if one really had an imaginary but water-like liquid such as was assumed to exist by Hamaker (1936) and Verwey and Overbeek (1948), the effects of its concomitant polar properties on immersed "lyophobic" particles would have a strong tendency to destabilize such particles through a solvophobic attraction. However, a compromise polar liquid could be ethanol. NaCl is only very slightly soluble in ethanol, but ethanol can accomodate up to a modest 0.05 M. KCl and it has a very low polar energy of cohesion (about $-5.2\ \text{mJ/m}^2$, as compared with $-102\ \text{mJ/m}^2$ for water; van Oss, 1994, 2006), so that its solvophobic effect would be rather small. Which probably is why ethanol has rarely, if ever, been the preferred liquid medium for the study of colloidal systems.

There really is only one realistic remedy to these problems and that is to make a proper study of the use of water as the liquid medium, not only because it is the liquid to which the electrical double layer part of the DLVO theory pertains most closely, but also because it is by far the most important liquid medium in most colloidal and in all biological systems. However, when using water as the liquid medium it is indispensable to extend the classical DLVO theory by including the *polar interaction energies* which always occur in water, using the Lewis acid–base approach. (Using the classical DLVO approach for an aqueous system limits one to content oneself with analyzing only about 10% of the forces that are actually operating in aqueous systems.)

## 2.2 Need for separate treatments of LW, AB and EL energies as a function of distance

Each one of the three different non-covalent forces involved in the extended DLVO (XDLVO) theory, i.e.: (1) Lifshitz–van der Waals (LW), or electrodynamic interactions; (2) Lewis acid-base (AB) or electron-acceptor/electron-donor interactions and: (3) electrical double layer (EL) interactions, decays according to fundamentally different regimens. This requires that for each one of the three types of interaction

**Table 3.1** Free energies ($\Delta G^{LW}$) and forces ($F^{LW}$) of unretarded[a] Lifshitz–van der Waals interactions[b] as a function of distance[c]

| Geometric configuration | $\Delta G^{LW}$ | $F^{LW}$ |
|---|---|---|
| Two parallel flat plates | $-A/12\pi(d-d_0)^2$ | $-A/12\pi(d-d_0)^3$ |
| Two spheres of equal radius, R[d] | $-AR/12(d-d_0)$ | $AR/12(d-d_0)^2$ |

a See Sub-section 3.2, below.

b Expressions for free energies ($\Delta G$) and forces (F) modified from van Oss (1994, 2006).

c The value for $d_0$ is discussed in Section 1.

d In the case of one sphere of radius, R and one flat plate, as well as with two cylinders of radius, R, crossed at right angles, both the $\Delta G$ and F values are twice the values of those given for the cases of two equal spheres.

the free energies as a function of distance be determined and treated separately. It is only after each energy vs distance curve thus obtained for LW, AB and EL interactions has been obtained separately, that all three (expressed in the same energy units) may be added together, to obtain the final XDLVO plot.

## 3. Decay with Distance of Lifshitz–van der Waals Interactions

### 3.1 LW decay with distance

In Table 3.1 the equations are given for the decay vs distance, d, of the LW free energies ($\Delta G^{LW}$) as well as of the LW forces ($F^{LW}$), for the geometric configurations of two plane parallel plates as well as for two equal spheres with radius, R. One can also find the values for the configuration of one flat plate and one sphere with radius, R, by multiplying the values for two equal spheres by a factor two. The macroscopic-scale Hamaker constant, $A_{ii}$, for a material or compound, i, is most easily derived from the $\gamma_i^{LW}$ value of that material or compound; see Eq. (3.4). On a macroscopic scale the Hamaker constant comprises all three van der Waals interactions, i.e., van der Waals–Keesom, or orientation forces, van der Waals–Debye, or induction forces, and van der Waals–London, or dispersion forces; see Chapter 1. Of these the van der Waals–London forces are quantitatively by far the predominant ones.

### 3.2 Retardation of the van der Waals–London forces

It should be noted that van der Waals–London interactions are, by virtue of the fact that they are caused by fluctuating dipolar molecules which in turn induce dipoles in other, neighboring molecules, subject to retardation. This is a consequence of the finite velocity of the propagation of dipole induction, even when this velocity is close to the speed of light. Thus, by the time the distance between the inducing dipole and the dipole that was to have been induced has been covered, the inducing dipole has already fluctuated to a somewhat less than perfectly complementary position vis-à-vis the molecule that was to have been induced as a dipole. This causes a lowering of the energy of attraction between the two. Thus, when $(d\text{–}d_0)$ approaches a value greater than 5 nm, the power to which $(d\text{–}d_0)$ is to be raised in the equations shown in Table 3.1, has to be increased by one. For example at $(d\text{–}d_0) > 5$ nm, for two parallel flat plates, $\Delta G^{LW}$ gradually becomes equal to $-A/12(d\text{–}d_0)^3$ and reaches exactly that value at approximately $d = 10$ nm and for two equal spheres $\Delta G^{LW}$ becomes $-AR/12(d\text{–}d_0)^2$ in a similar manner. $F^{LW}$ has to be treated similarly, from $(d\text{–}d_0) > 5$ nm on. This phenomenon of retardation of van der Waals–London interactions was first described by Casimir and Polder (1948), prompted by a suggestion from J.Th.G. Overbeek. The existence of the retardation effect was ultimately experimentally confirmed by Tabor and Winterton (1969).

Given the predominance of van der Waals–London forces among the total of all three varieties of Lifshitz–van der Waals interactions, it is always necessary to take the influence of van der Waals–London retardation into account at $d \geqslant 10$ nm. It has not yet been experimentally determined whether for values of $d > 10$ nm the retardation effect of $\Delta G^{LW}$ persists, or changes, or disappears. This is because at values for d greater than about 50 to 100 nm, $\Delta G^{LW}$ becomes exceedingly small at these distances; see also Israelachvili (1985, 1991). However, in any case, after the first 3 to 5 nm LW energies decay very gradually, but LW energies are rarely large to begin with, so that a further gradual LW decay beyond a distance, $d > 5$ to 10 nm (where retardation begins to set in) seldom seriously impacts the total outcome of, e.g., particle stability in aqueous systems.

## 3.3 Attractive and repulsive LW interactions

In the $\Delta G_{SWS}{}^{LW}$ mode, i.e., pertaining to the LW interaction energy between two *similar* molecules or particles, S, immersed in water, the Lifshitz–van der Waals component of the interaction is always attractive, i.e., $\Delta G_{SWS}{}^{LW}$ is always negative. However, as is already mentioned in Chapter 1, when two *different* molecules or particles, S′ and S″, interact, $\Delta G_{S'WS''}{}^{LW}$ can be positive (i.e., when $\gamma_{S'}{}^{LW} > \gamma_W{}^{LW} > \gamma_{S''}{}^{LW}$, or when $\gamma_{S'}{}^{LW} < \gamma_W{}^{LW} < \gamma_{S''}{}^{LW}$), because $\gamma_i{}^{LW}$ is proportional to the Hamaker constant of material, i, that is, to $A_{ii}$; see Chapter 1 and see also Sub-section 1.4 of this chapter, see above. In actual practice a Lifshitz–van der Waals repulsion in water occurs with some frequency when material S′, immersed in water, is a perfectly common material, such as glucose, or poly(methylmethacrylate) (PMMA), but material S″ is the water–air interface (a), whose $\gamma_a{}^{LW}$ is zero. In such cases (see Chapter 1), $\Delta G_{swa}{}^{LW} > 0$ (see, e.g., Chaudhury and Good, 1983, who, however, did not yet take AB interactions into account, mainly because they wrote the 1985 paper before the AB approach was sufficiently developed for ready application at that date). For instance, in the case of glucose, $\Delta G_{swa}{}^{LW} = +34.2$ mJ/m$^2$, but glucose, dissolved in water is not only repelled by the water–air interface via repulsive LW forces, but also by repulsive $\Delta G_{swa}{}^{AB}$ forces of an additional $+50.8$ mJ/m$^2$, to achieve a total repulsion by the water air interface of $+85$ mJ/m$^2$.[1] Also, with PMMA, $\Delta G_{swa}{}^{LW} = +5.8$ mJ/m$^2$, but there is, in addition, a $\Delta G_{swa}{}^{AB}$ *attraction* of $-67$ mJ/m$^2$, so that the LW repulsion (being less than 10% of the total attraction of $-61.2$ mJ/m$^2$) remains virtually unnoticed.[2] Thus, as the absolute value of $\Delta G_{S'WS''}{}^{LW}$ is almost always smaller than $\Delta G_{S'WS''}{}^{AB}$, a perfectly possible van der Waals repulsion between two different entities, immersed in water, usually remains unnoticed, especially in those cases where S″ represents a water–air interface. However, in organic liquids such as benzene, toluene, tertrahydrofuran, carbon tetrachloride, chlorobenzene and chloroform, *exclusively LW-driven phase separations* between pairs of polymers, each dissolved in the same organic liquid [e.g., pairs formed among poly(isobutylene), PMMA, polystyrene, polyvinyl chloride] is perfectly feasible (van Oss, Chaudhury and Good, 1989; see also van Oss, 1994, 2006).

### 3.4 Mechanism of LW action at a distance in water

Lifshitz–van der Waals interactions play two separate roles in the action at a distance in water. The first one is a general role, inherent in the electrodynamic nature of all three van der Waals forces (see, e.g., Israelachvili, 1985, 1991). This is also the role described in the energy vs distance equations given in Table 3.1. The second role of LW forces lies in the propagation of attractions due to hydrophobic (AB-driven) interactions in water, insofar that hydrophobic particles, molecules or surfaces become hydrated in water by purely LW forces. The first (closest) resulting hydrated surface layer becomes in its turn hydrophobic and the second layer of hydration becomes somewhat less hydrophobic, etc. In this manner the LW-driven hydration of hydrophobic surfaces in water causes a hydrophobic effect-driven (i.e., AB-driven) attraction at a distance; see Chapter 10, Section 2.

## 4. Decay with Distance of Lewis Acid–Base Interactions

### 4.1 Decay with distance of AB free energies and forces

Table 3.2, which shows the free energies ($\Delta G^{AB}$) and forces ($\Delta F^{AB}$) as a function of distance, d, for the same geometric configurations as also used for LW interactions (see Table 3.1). In contrast with the LW energies and forces, Lewis acid–base energies and forces decay exponentially, with a characteristic length ($\lambda$), or ("decay length") for water, such that $\lambda = 0.6$ nm, at 20 °C. The decay length of water at $\lambda \approx 0.6$ nm corresponds to the radius of gyration, $R_g$, of clustered water, but at the molecular weight of water at 18, the value of $R_g$ should be about 0.2 nm (Chan et al., 1979; Parsegian et al., 1979). However, it should not be overlooked that at a temperature of about 20 °C, water occurs on average as clusters of the size of approximately 4.5 molecules (see Chapter 9, Section 3), corresponding to a cluster molecular weight of 81, which correlates more closely with an $R_g$ value of $\lambda \approx 0.6$ nm. It may be assumed that at higher temperatures $R_g$ and concomitantly, $\lambda$, decreases to about 0.2 nm when reaching the boiling point of water; see Chapter 9, Sub-section 1.2. The characteristic length, $\lambda$ of clustered water, which

[2] The surface tension properties of poly(methylmethacrylate), or PMMA, are: $\gamma^{LW} = 40.6$, $\gamma^{+} = 0$, $\gamma^{-} = 12.0$; all in mJ/m$^2$ (van Oss, Chaudhury and Good, 1989). See Eq. (2.13) for obtaining $\Delta G_{132}$, or in this case, $\Delta G_{swa}$.

**Table 3.2** Free energies ($\Delta G^{AB-}$) and forces ($F^{AB}$) of Lewis acid–base interactions, as a function of distance, d[a]

| Geometric configuration | $\Delta G_d{}^{ABd}$ | $F_d{}^{ABd}$ |
|---|---|---|
| Two parallel flat plates | $\Delta G^{AB''} \exp[(d_0-d)/\lambda]^b$ | $-\Delta G^{AB''} \exp[(d_0-d)/\lambda]1/\lambda^b$ |
| Two spheres of equal radius, R[c] | $\pi R\lambda\Delta G^{AB''} \exp[(d_0-d)/\lambda]^b$ | $-\pi R\Delta G^{AB''} \exp[(d_0-d)/\lambda]^b$ |

*a* Expressions for free energies ($\Delta G$) and forces (F) modified from van Oss (1994, 2006).

*b* The superscript ($''$) indicates that $\Delta G''$ pertains to the plane parallel plates configuration, *at the minimum equilibrium distance*, $d_0$; $\lambda$ is the characteristic, "decay" length of water; $\lambda \approx 0.6$ nm, see Chapter 9, Section 3.

*c* In the case of one sphere of radius, R and one flat plate, as well as with two cylinders of radius, R, crossed at right angles, both the $\Delta G$ and F values are twice the values of those found for two equal spheres.

*d* The value for $d_0$ is equal to the one discussed in Section 1.

at 20 °C has been shown to have to be shorter than 1.0 nm, has been estimated, from considerations on the hemagglutination of human erythrocytes with IgG isotype antibodies, to be closest to 0.6 nm, which also correlates best with a water cluster size of ≈4.5 water molecules per cluster (van Oss, 1990, 1994, pp. 318–327; 2006, pp. 336–339); see also Chapter 9, Section 3.

The exponential rate of decay vs distance, d, causes the usually rather strong AB interactions in water (given a decay length of $\lambda \approx 0.6$ nm) to make themselves felt up to distances, $d \leqslant 8$ nm in the attractive as well as in the repulsive mode. At distances beyond about 8 nm they decay to negligible values. In general, in polar systems, as a consequence of the high $\Delta G^{AB}$ values close to $d = d_0$, AB energies, be they attractive or repulsive, usually play the dominant role in the outcome of hydrophobic attractions as well as of hydrophilic repulsions *in water* (see, e.g., Chapter 8).

## 4.2 Mechanisms of AB attractions and repulsions at a distance in water

Both in hydrophobic attractive (AB) and in hydrophilic repulsive (AB) interactions in water, *hydration* plays the major role. Contrary to hydrophobic attraction, which is caused by AB forces, hydrophobic *hydration* is mainly caused by LW forces. Thus, curiously, LW-caused *hydrophobic hydration* is the initial driving force for the propagation at a distance of hydrophobic (AB) *attractions* in water; see Sub-section 3.4, above, as well as Chapter 10, Section 2.

AB-driven hydrophilic *repulsion* at a distance, in water, is also driven by hydration, but in this case the hydration is mainly (i.e., for 50% or even somewhat more) caused by AB forces, in such a way that hydrophilic hydration acts in orienting the water dipoles with their H atoms toward the hydrophilic surface, thus extending the predominant electron-donicity of the hydrophilic particles' surfaces outward into the aqueous medium; see Chapter 10, Section 3; see also Chapter 5, Eq. (5.1), term 5.

**Table 3.3** Free energies ($\Delta G^{EL}$) and forces ($F^{EL}$) of electrical double layer interactions, as a function of distance, d, for relatively Weak Interactions, i.e., for $|\zeta| \leqslant 50$ mV[a]

| Geometric configuration | $\Delta G^{EL}$ | $F^{EL}$ |
|---|---|---|
| Two parallel flat plates | $1/\kappa 64\pi kT\gamma_0^2 \exp[\kappa(d_0-d)]^b$ | $-64\pi kT\gamma_0^2 \exp[\kappa(d_0-d)]^b$ |
| Two spheres of equal radius, R[c] | $0.5\varepsilon R\psi_0^2 \ln[1+\exp\{\kappa(d_0-d)\}]^d$ | $-0.5\varepsilon R\psi_0^2 \ln[1+\exp\{\kappa(d_0-d)\}]\kappa^d$ |

a Expressions for free energies ($\Delta G$) and forces (F) modified from van Oss (1994, 2006).

b $\gamma_0 = [\exp(\nu e\psi_0/2kT) - 1]/[\exp(\nu e\psi_0/2kT) + 1]$; n = the number of ions of each species per $cm^3$ of liquid; k is Boltzmann's constant, i.e., $= 1.38 \times 10^{-23}$ J $deg^{-1}$ and T is the absolute temperature in degrees Kelvin.

c In the case of one sphere of radius, R, and one flat plate, as well as with two cylinders of radius, R, crossed at right angles, both the $\Delta G$ and F values are twice the values of those found for two equal spheres.

d $\varepsilon$ is the dielectric constant of the liquid; for water, $\varepsilon \approx 80$; $\psi_0$ is the electric potential of the surface of the charged entity (see Sub-section 3.1); $1/\kappa$ is the thickness of the diffuse electrical (Debye) double layer (see Sub-section 3.1).

# 5. Decay with Distance of Electrical Double Layer Interactions

## 5.1 Equations and relation between the $\zeta$-potential and the $\psi_0$-potential

Table 3.3 gives the equations for finding the free energies ($\Delta G^{EL}$) and forces ($F^{EL}$) involved in EL interactions as a function of distance, d. It should be noted that whilst LW and AB free energies for a condensed-phase material can be determined together via contact angle measurements on such a material, using different appropriate contact angle liquids (see Chapter 2, Sub-section 1.6), EL free energies have to be determined via one of the electrokinetic methods (cf. van Oss, 1994, 2006), i.e., most commonly, electrophoresis. However, such methods only allow one to determine the *$\zeta$-potential* of (electrically charged) particles or macromolecules, which is the potential at the slipping plane between the hydrated particle (where the hydration layer also contains some of the electrolytes used in the aqueous medium) which is being transported in the electric field that makes the particle move with respect to the aqueous medium. On the other hand, what one needs to obtain the EL free energies of charged particles is the $\psi_0$-potential at the very surface of these particles, *inside* the layer of water of hydration and included salt ions that move with the particle. For $\zeta$-potentials smaller than about 50 mV, the connection between the *measured* $\zeta$-potential and the desired $\psi_0$-potential is:

$$\psi_0 = \zeta(1 + z/R)e^{\kappa z} \tag{3.5}$$

(van Oss, 1994, 2006); where z is the thickness of the hydration layer traveling with the particle, when driven by the electric field (z usually has a thickness between 0.3 and 0.5 nm), R is the radius of the moving particle and $1/\kappa$ is the thickness of the diffuse ionic double layer, or Debye length:

$$1/\kappa = [(\varepsilon kT)/(4\pi e^2 \Sigma v_i^2 n_i)]^{0.5}, \tag{3.6}$$

where $\varepsilon$ is the dielectric constant of the liquid (for water, $\varepsilon = 80$); for kT, see Table 3.3 (at 20 °C, 1 kT $= 4.04 \times 10^{-14}$ ergs $= 4.04 \times 10^{-21}$ J); $v_i$ is the valency of each ion species of the salt(s) dissolved in the aqueous medium and $n_i$ is the number of ions of each species per $cm^3$ of water. For instance, for a $\mu = 0.15$ (isotonic) NaCl solution, $1/\kappa = 0.8$ nm. For a list of $1/\kappa$ values for different salts, see van Oss (1994, 2006).

## 5.2 Electrokinetic determination of $\zeta$-potentials

By far the most utilized electrokinetic method for determining $\zeta$-potentials is electrophoresis. To obtain $\zeta$-potentials of, e.g., particles, or polymers (such as proteins or polysaccharides), one measures their electrophoretic mobility. With small particles or cells this done by microelectrophoresis and with dissolved biopolymers via electrophoresis in gels or porous membranes (van Oss, 1994, 2006). For descriptions of the various most common methodologies as well as for general principles of electrophoresis, see also Shaw (1969) and Hunter (1981) and for cell electrophoresis see Bauer (1994).

From the electrophoretic mobility (u) measured by electrophoresis and expressed in units of $\mu V^{-1} s^{-1}$ cm (i.e., from measured electrophoretic velocities in µm/s per volt per cm of the direct current electric field used), the particles' or cells' or macromolecules' $\zeta$-potential can be derived, choosing one of two possible equations dependent upon one of the following conditions:

### 5.2.1 Thick ionic double layer

For particles with radius, R, immersed in an aqueous medium containing low concentrations of dissolved salts such that the Debye length, $1/\kappa$, is much greater than R, i.e., at low R values (e.g., $\kappa R < 0.1$), one uses the Debye–Hückel equation:

$$\zeta = (6\pi\eta u)/\varepsilon, \tag{3.7}$$

where $\eta$ is the viscosity of the aqueous medium and $\varepsilon$ (see also Eq. (3.6)) is the dielectric constant (permittivity) of water (Overbeek, 1952; see also van Oss, 1994, 2006).

### 5.2.2 Thin ionic double layer

For particles with radius, R, immersed in an aqueous medium containing relatively high concentrations of dissolved salts, such that the Debye length, $1/\kappa$, is much smaller than R, in other words, at high R values (i.e., $\kappa R > 300$) one uses von Smoluchowski's equation:

$$\zeta = (4\pi\eta u)/\varepsilon \tag{3.8}$$

(Overbeek, 1952; see also van Oss, 1994, 2006). For $\kappa R$ values which are intermediate between 0.1 and 300 one uses numerical factors intermediate between 6 (Eq. (3.7)) and 4 (Eq. (3.8)) (Henry, 1931; see also Overbeek, 1952).

In most cases, when the electrophoretic mobility, u, is known and for $\zeta$-potentials which are not greater than about 50 mV, $\zeta$ can be found with Eq. (3.7) or Eq. (3.8). However, when $\zeta$ is significantly greater than 50 mV, and at $\kappa R$ values that are intermediate between 0.1 and 300, a relaxation effect involving the diffuse electrical double layer sets in, which causes a retardation of the electrophoretic mobility when one has electrical double layers that are neither very thin and dense, nor very wide and unsubstantial (Overbeek, 1943; Overbeek and Wiersema, 1952). In the above-mentioned ranges of high $\zeta$-potentials and intermediate $\kappa R$ values the relaxation effect causes a highly significant decrease in electrophoretic mobility. However in such cases the correct $\zeta$-potential can be found by consulting the appropriate $\zeta$-potential vs $\kappa R$ graphs given by Overbeek and Wiersema (1952); see also van Oss (1994, 2006).

Once the value for $\zeta$ is found by electrophoretic measurements, $\psi_0$ can be found in most cases via Eq. (3.5) after which the appropriate $\Delta G_{EL}$ vs distance (d) equations can be found in Table 3.3.

# 6. Influence of the Ionic Strength on Non-Covalent Interactions in Water

## 6.1 Definition of ionic strength

The ionic strength (μ) of (usually aqueous) salt solution is expressed as:

$$\mu = 0.5(\Sigma v_i^2 C_i) \qquad (3.9)$$

where $v_i$ is the valency of each ion (see also Eq. (3.6)) and $C_i$ its concentration. For instance, for 1 M NaCl, $\mu = 1.0$, whilst for 1 M $Na_2SO_4$, $\mu = (2^2 \times 1.0 + 2 \times 1.0)/2 = 6/2 = 3.0$ (see also van Oss, 1994, 2006, for further examples).

## 6.2 Influence of ionic strength on LW interactions

The influence of moderate amounts of, e.g., $Ca^{2+}$ ions on the $\gamma^{LW}$ of negatively charged particles (by means of added $CaCl_2$), which causes a decrease in the $\zeta$-potential as well as in the $\gamma^-$ value of these particles, has been shown to be practically zero (Wu et al., 1994a, 1994b). At moderate ionic strengths there would, in any event, be no obvious reason why the addition of neutral salts (at lower concentrations than those needed to cause "salting out") would significantly influence LW interactions. At high ionic strengths, however, the counterions might fairly densely coat an oppositely charged particle or macromolecule, so that it would be measured as having the LW properties of that counterion, instead of the original LW properties of the particle or macromolecule.

## 6.3 Influence of the ionic strength on AB interactions

There are two ways by which added salt ions can influence AB interactions:

### 6.3.1 Through the influence of plurivalent counterions on oppositely charged particles or macromolecules

Low concentrations of plurivalent counterions will *lower* the $\zeta$-potential of such particles or molecules, which concomitantly also causes a decrease in their $\gamma^-$ parameters via an amplifying EL–AB linkage, which makes the particles or molecules less hydrophilic, or changes them from hydrophilic to hydrophobic. For example the addition of 0.47 mM $LaCl_3$ ($\mu = 0.0028$) will cause glass particles in a stable aqueous suspension with an initial $\gamma^-$ of 31.7 mJ/m$^2$ to change to a $\gamma^-$ of 20.9 mJ/m$^2$, which turns these particles' surfaces from hydrophilic (i.e., mutually repulsive when immersed in water) to hydrophobic (i.e., mutually attractive when immersed in water), which causes them to destabilize (i.e., to flocculate) from their initial stable aqueous suspension (Wu, 1994; see also Chapter 8, Section 3).

### 6.3.2 Through the hydrophobizing influence of high salt concentrations

High salt, as well as high sugar concentrations (Docoslis et al., 2000) *increase* the AB-driven free energy of cohesion of water, which in turn increases the hydrophobizing effect of the aqueous solution, so that marginally hydrophilic particles, which are in a stable aqueous suspension, or marginally soluble macromolecules, respectively flocculate, or precipitate. With high salt concentrations this phenomenon is called "salting out"; see Chapter 5 Sub-section 2.2.1.

Finally, at lower than the high salt concentrations alluded to under Sub-section 6.3.2, above, but at ionic strengths, $\mu \approx 0.10$ or higher (e.g., at the physiological $\mu = 0.15$ of human blood serum), $\Delta G^{AB}$ is very little affected whilst $\Delta G^{EL}$ becomes negligible, due to its very steep decay with distance at these higher ionic strengths. Therefore, in most mammalian systems, $\Delta G^{AB}$ is the major source of hydrophilic repulsion, which assures blood cell stability as well as protein solubility in biological liquids; see also Chapter 9, Section 3.

## 6.4 Influence of the ionic strength on EL interactions

Electrical double layer-driven (EL) interactions have long been known to be strongly influenced by the ionic strength of the aqueous medium, see for instance the influence of the Debye length ($1/\kappa$) on all $\Delta G^{EL}$ vs d equations given in Table 3.3.

Equation (3.6) shows the inclusion of the ionic strength in the expression for $1/\kappa$, where $n_i$ is N/1000 times greater than $C_i$ (Eq. (3.9)), N being Avogadro's number, i.e., the number of molecules per mol $\approx 6 \times 10^{23}$.

At very low ionic strengths the decay of $\Delta G^{EL}$ vs d is exceedingly gradual, i.e., it shows as a line which is almost parallel to the abscissa (on which the distance, d, is plotted), as a consequence of an extremely long Debye length ($1/\kappa$). On the other hand, at high ionic strengths, $1/\kappa$ is quite small, decreasing to 1 nm or less, causing $\Delta G_{EL}$ to decline very steeply as a function of d. Thus at low ionic strengths

$\Delta G^{EL}$ can be of crucial importance in assuring the stability of aqueous suspensions of charged particles (cf. Chapter 8). On the other hand in mammalian biological systems, whose aqueous media have a high ionic strength (in humans $\mu = 0.15$), the importance of EL forces in assuring cell stability (Chapter 8) or biopolymer solubility (Chapter 7) is quite minor.

## 7. An EL–AB Linkage

There would seem to exist an EL–AB linkage. Surfaces with an elevated $\zeta$-potential (i.e., $|\zeta| \geqslant 50$ mV) tend to be accompanied by a high $\gamma^-$ parameter (e.g., $\gamma^- \geqslant 30$ mJ/m$^2$; Wu, 1994; Wu et al., 1994a, 1994b). Now, when one adds small amounts of salts with plurivalent counterions (e.g., cations) to such electrically charged (here negatively charged) surfaces, through the admixture of plurivalent cations (which are electron-acceptors), the $\zeta$-potential, and with it, the $\gamma^-$ parameter becomes partly neutralized so that it decreases strongly in value, which typically causes such initially strongly hydrophilic electron-donating surfaces to become *hydrophobic*. This EL–AB connection appears to act exclusively on electrically charged surfaces.

Dr. A. Dobry (1948) was probably among the first to show the occurrence of complex coacervation, or even flocculation, between Lewis basic and Lewis acidic polymers, immersed in water. It was clear that such polymer pairs only form complexes together under conditions where polymer 1 is positively and polymer 2 negatively charged when immersed in water. [It should be noted that although Mme. Dobry did quote Bungenberg de Jong (1949) from a French translation), in her paper of 1948 she did not clearly distinguish between *coacervation* (the separation between two different dissolved polymers into two phases, through mutual *repulsion*) and *complex-coacervation* (the *attraction* between two different polymers, leaving a polymer-rich and a polymer-depleted phase), which makes her description of the results of both phenomena somewhat confusing. However upon further scrutiny, her results leave no doubt.]

On the other hand very hydrophilic materials such as dextran, which are *neutral*, but have an elevated $\gamma^-$ parameter (i.e., 55 mJ/m$^2$; see van Oss, 1994, 2006) and a close-to-zero $\zeta$-potential (van Oss, Fike et al., 1974), does *not* undergo changes in its $\gamma^-$ parameter by the addition of moderate amounts of salts (at concentrations below those required for salting-out; i.e., below concentrations higher than 1 molar) to the aqueous medium, nor by altering that medium's pH. Another example of a very hydrophilic, strongly electron-donating polymer with a zero (or negligible) $\zeta$-potential is polyethylene oxide (PEO), also often called polyethylene glycol (PEG) (see van Oss, 1994, 2006). To summarize: changes in the value of the $\zeta$-potential can accompany changes in the value of $\gamma^-$, but the value of $\gamma^-$ cannot readily be influenced in this manner when the material has no significant $\zeta$-potential; see also Chapter 8, Section 3, as well as Chapter 12, Sub-sections 1.3 and 3.3.

# 8. Role of the Radius of Curvature, R, of Round Particles or Processes in Surmounting AB Repulsions in Water

It will be apparent from all three tables (Tables 3.1–3.3) that in the case of all three non-covalent forces (LW, AB and EL), applied to spherical objects, particles, cells or molecules, the free energy of interaction at any distance, d, is directly proportional to the radius of curvature, R, of the object in question. This fact allows us to explain how, when immersed in an aqueous medium, no mammalian blood cells, such as red cells or leukocytes or platelets, will be able to approach and attach to vessel walls or to each other *in vivo*, as long as they remain approximately spherical, because they are normally fairly smooth, with a relatively large radius of curvature. There are however situations when blood cells such as phagocytic neutrophils (which are first line of defense leukocytes), need to approach and subsequently adhere to, internalize and eliminate invading bacterial cells. When such neutrophils receive chemical ("chemotactic") signals of the presence of bacterial cells in their vicinity, they move more closely to the invaders and start protruding long arm-like *cylindrical processes with a small radius of curvature* and aim these at the bacteria closest to them. These cylindrical processes with a small radius of curvature can pierce the normally existing repulsion field owing to their small R value. After contact is made with a bacterial cell by means of such a long thin protrusion, adhesion occurs, followed by endocytosis of the bacterium into the long tube, to be steered toward a phagosome inside the main body of the neutrophil, where it is killed and eliminated by phagosomal $H_2O_2$, plus enzymes (see van Oss et al., 1975).

Similarly, circulating blood platelets are normally oval and smooth, but when activated by adenosine phosphate (ADP), triggered by the occurrence of a bleeding episode in the vicinity, platelets become "sticky." More precisely this actually means that they become *spiculated*, so that their newly formed tiny spikes, thanks to their small radius of curvature, can penetrate the repulsion fields which keep normal, smooth platelets apart, and thus become enabled to make contact and form a clot which closes the hole where the bleeding originated (White, 1968; van Oss et al., 1975).

Another example of the role of thin, pointy protrusions with a small radius of curvature on otherwise smooth, quasi-spherical bodies is that of pathogenic viruses. Many of these have close-to-spherical bodies, but with most of them the central spheres are surrounded by long thin processes with a small radius of curvature, which are distally provided with a ligand with which they able to attach themselves to a receptor on the cell which they aim to invade. (It should be noted that viruses cannot reproduce outside of living cells). Here also the viral protrusions with a small radius of curvature serve to overcome the repulsion field of mammalian (e.g., blood) cells, while the distal ligands of these thin processes serve to create a bond with the cell's receptors, upon which interiorizing of the virus follows.

A schematic presentation of the role of thin cylindrical processes with a small radius of curvature, protruding from an otherwise smooth sphere (with a large radius of curvature) is given in Figure 3.1, where it is made clear how such thin cylindri-

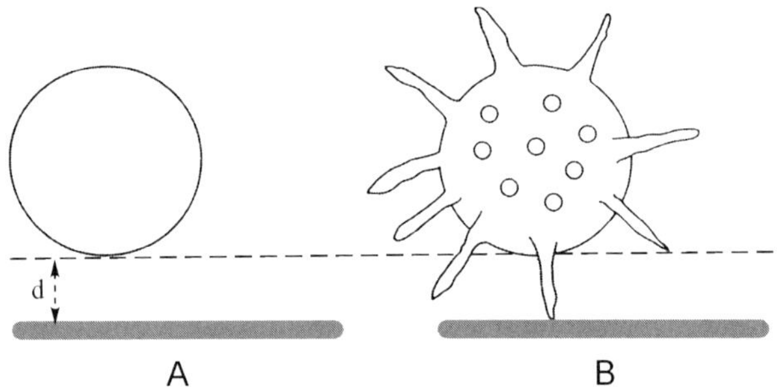


**Figure 3.1** Illustration of the differences occurring at conditions where repulsion fields exist *in vivo* between hydrophilic spherical bodies among one another as well as between such spherical bodies and other hydrophilic surfaces, immersed in water, showing a smooth sphere with a large radius of curvature (left), which cannot penetrate a repulsion field beyond a distance, d, in contrast with a similar sphere endowed with long, thin cylindrical protrusions with small radii of curvature (right). The latter can easily penetrate the repulsion field thanks to the protruding cylindrical processes with a small radius of curvature. For the numerical role of the radii of curvature (R) see Tables 3.1–3.3, in all cases of interactions between two spheres of radius R and of interactions between a sphere of radius R and a flat surface endowed with thin protrusions (illustrated above). From van Oss, (2003, pp. 177–190), with permission from Wiley–Interscience, New York, NY.

cal protuberances can pierce repulsion fields emanating from other large bodies or surfaces, which large, smooth spheres cannot penetrate.

## 9. Comparison between Direct Measurements via Force Balance or Atomic Force Microscopy, and Data Obtained via Contact Angle Determinations, in the Interpretation of Free Energies vs Distance Plots of the Extended DLVO Approach

### 9.1 Direct measurements of forces vs distance

With Israelachvili's force balance (see, e.g., Israelachvili, 1985, 1991), or via an Atomic Force Balance type of approach, one only obtains the final total force (or energy) vs distance curve. Forces are readily transformed into free energies as a function of distance, d, using Derjaguin's (1934) approximation:

$$\Delta G_d = -F_d/2\pi R \quad \text{(for two crossed cylinders, as well as for a sphere and a flat plate)} \qquad (3.10A)$$

or:

$$\Delta G_d = -F_d/\pi R \quad \text{(for two equal spheres).} \qquad (3.10B)$$

However, with Extended DLVO plots obtained via direct measurements by means of a Force Balance or an Atomic Force Microscopy device one can only obtain the *total* free energy vs distance curve, without gaining further information concerning the separate Lifshitz–van der Waals, Lewis acid-base or electrical double layer contributions.

## 9.2 Determination of the separate LW, AB and EL contributions

### 9.2.1 The LW contributions

Macroscopic-scale Lifshitz–van der Waals forces represent one single LW property of material, i, ($\gamma_i^{LW}$), often used in conjunction with the single (LW) property of the aqueous medium, w, ($\gamma_w^{LW}$); see term 1 of Eq. (5.1). Here, $\gamma_w^{LW}$ is well known in advance ($\gamma_w^{LW} = 21.8$ mJ/m$^2$ at 20 °C), but $\gamma_i^{LW}$ should in each case be measured by contact angle determination with, e.g., drops of diiodomethane deposited on material, i. When acting between two identical entities (i), $\Delta G_{iwi}^{LW}$ is always negative, i.e., attractive [see the LW part (term 1) of Eq. (5.1)], or zero.

### 9.2.2 The AB contributions

AB contributions are more complex than either LW or EL interactions. Between two *identical* entities (i), immersed in water, $\Delta G_{iwi}^{EL}$ *is always positive, i.e., repulsive,* whilst $\Delta G_{iwi}^{LW}$ *is always negative, i.e., attractive (or zero).* However, free energies of interaction between two identical entities i, immersed in water and expressed as $\Delta G_{iwi}^{AB}$, *can be positive (i.e., repulsive), or zero (i.e., neutral) or negative (i.e., attractive).* For instance, in the case of non-polar particles, immersed in water, $\Delta G_{iwi}^{AB} = -102$ mJ/m$^2$, i.e., strongly attractive. This attractive energy (cf. term 3 of Eq. (5.1) in Chapter 5) is always present: it is the term defining the hydrophobic effect; see Chapter 5, Sub-section 2.1. That is the reason why $\Delta G_{iwi}^{AB}$ can only turn positive (i.e., repulsive) when (usually) term 5 of Eq. (5.1) becomes more positive than the $+102$ mJ/m$^2$ needed to surmount the always-present hydrophobic attraction caused by the AB free energy cohesion of water (i.e., term 3 of Eq. (5.1)).

In practice, only contact angle determinations, using several liquids, can tell whether $\Delta G_{iwi}^{AB} > 0$, $\Delta G_{iwi}^{AB} = 0$, or $\Delta G_{iwi}^{AB} < 0$, and by how much. There is no direct mechanical measurement device (such as a Force Balance) which allows one to find the sign and precise value of $\Delta G_{iwi}^{AB}$, although Force Balance results that were either too repulsive or too attractive to be caused by the interplay between van der Waals attractions and electrical double layer repulsions normally found with the classical DLVO approach have experimentally divulged the existence of non-classical DLVO interactions [Israelachvili and Pashley (1984); Pashley et al. (1985); Claesson (1986) and more recently Grasso et al. (2002)]. Thus, at the present state of the art, AB as well as LW interaction energies still are best determined via contact angle measurements, see Chapter 4, Section 3, below.

### 9.2.3 The EL contributions

The theory of electrical double layer interaction energies (which between identical particles (i), immersed in water, are always positive (i.e., repulsive) has been well established since Hűckel (1924) and von Smoluchowski (1921), who each studied somewhat different aspects of EL interaction energies and where each studied a different set of conditions of thickness of the diffuse electrical double layer ($1/\kappa$)

to particle radius (R) ratio ($\kappa R$). The different contingencies described by these two Authors pertain to different experimental conditions, both of which occurring with approximately equal frequency, see Sub-section 5.2, above.

## 9.3 Advantages and disadvantages of the extended DLVO approach vs contact angle determinations

### 9.3.1 Using a force balance

One measures rather accurately (using, e.g., a Force Balance) the total aggregate value of $\Delta G_{iwi}^{TOT} (= \Delta G_{iwi}^{LW} + \Delta G_{iwi}^{AB} + \Delta G_{iwi}^{EL})$ at distances $d > d_0$. However, at $d = d_0$, the direct measurement of $\Delta G_{iwi}^{TOT}$ can be somewhat less accurate for mechanical reasons, as a consequence of some degree of distortion occurring at "contact."

### 9.3.2 Contact angle measurements

Contact angle measurements tend to yield the most accurate values for $\Delta G_{iwi}^{LW}$ and $\Delta G_{iwi}^{AB}$ at $d = d_0$, i.e., *at "contact."* However, at $d > d_0$ one depends on calculations by Extended DLVO rules (see Tables 3.1, 3.2, 3.3), which is more laborious and therefore more prone to mistakes so that this approach may become less accurate than direct measurements for distances greater than $d_0$.

SECTION B

# Surface Thermodynamic Properties of Water with Respect to Condensed-Phase Materials Immersed in It

CHAPTER FOUR

# Determination of Interfacial Tensions between Water and Other Condensed-Phase Materials

## Contents



## 1. The Interfacial Tension between a Solid (S) and a Liquid (L)

### 1.1 Importance of the interfacial tension ($\gamma_{SL}$) between S and L

The interfacial tension ($\gamma_{SL}$) between a solid (S) (or other condensed-phase material) and a liquid (L) is a term of prime importance as a part of all versions of the Dupré equations that govern the interaction energies between condensed-phase materials and the liquid in which they are immersed (see Chapter 2). [A solid (S) may also take the place of another liquid that is immiscible with liquid (L), or even of a solute that is soluble in liquid (L).]

*Interface Science and Technology*, Volume 16
ISSN 1573-4285, doi: 10.1016/S1573-4285(08)00204-4

In addition it should be noted that $\gamma_{SL}$ is proportional (by a factor of $-2$) to the free energy of interaction between two particles or molecules of S, immersed in liquid, L, i.e., $\Delta G_{SLS}$ (see Eq. (2.11)). In the next chapter (Chapter 5) the role of $\Delta G_{SLS}$ and $\Delta G_{SWS}$ [where the liquid, L, is taken to be water (W)] and thus also of $\gamma_{SL}$, $\gamma_{SW}$ or $\gamma_{iw}$, is described in treatments of the two opposing phenomena of hydrophobic attraction forces (the "hydrophobic effect") and of hydrophilic repulsion forces ("hydration pressure"). Both of these two non-covalent interactions occurring in water play unique roles as the two opposing Lewis acid–base-driven forces that dominate in all aqueous colloidal and biological systems.

### 1.2 The polar versions of $\gamma_{SL}$ and $\gamma_{iw}$

The most general versions of the equations describing interaction energies are the polar versions of the Dupré equations (see Eqs. (2.9), (2.11), (2.13)), as well as of the Young equation (Eq. (2.14)) and the Young–Dupré equation (Eq. (2.15B)). The polar version of the $\gamma_{SL}$ equation is repeated here (see Eq. (2.6)):

$$\gamma_{SL} = (\sqrt{\gamma_S{}^{LW}} - \sqrt{\gamma_L{}^{LW}})^2 + 2[\sqrt{(\gamma_S{}^{+}\gamma_S{}^{-})} + \sqrt{(\gamma_L{}^{+}\gamma_L{}^{-})} - \sqrt{(\gamma_S{}^{+}\gamma_L{}^{-})} - \sqrt{(\gamma_S{}^{-}\gamma_L{}^{+})}]. \quad (4.1)$$

Because of the exceptional importance of water among polar liquids, the polar version of the interfacial tension between a polar condensed-phase material (i) and water (w) is also given:

$$\gamma_{iw} = (\sqrt{\gamma_i{}^{LW}} - \sqrt{\gamma_w{}^{LW}})^2 + 2[\sqrt{(\gamma_i{}^{+}\gamma_i{}^{-})} + \sqrt{(\gamma_w{}^{+}\gamma_w{}^{-})} - \sqrt{(\gamma_i{}^{+}\gamma_w{}^{-})} - \sqrt{(\gamma_i{}^{-}\gamma_w{}^{+})}]. \quad (4.2)$$

## 2. The Interfacial Tension between an *Apolar* Material or Compound (A) and Water (W)

### 2.1 The $\gamma_{AW}$ equation

When i (Eq. (4.2)) is completely apolar, i may be designated as A and the interfacial tension with water as $\gamma_{AW}$, so that:

$$\gamma_{AW} = (\sqrt{\gamma_A{}^{LW}} - \sqrt{\gamma_W{}^{LW}})^2 + 2\sqrt{(\gamma_W{}^{+}\gamma_W{}^{-})}, \quad (4.3)$$

where it will be noted that all *polar* terms of Eq. (4.2) containing the subscript (A) instead of (i) have disappeared because A is apolar and therefore can neither combine in a polar manner with itself nor with water ("apolar" is another term for "non-polar"). It should also be noted that LW interaction energies between an apolar material, A, and water, W, are not affected by the apolarity of A and that the self-hydrogen-bonding term for water also remains in place in Eq. (4.3).

Thus, for a typical apolar material such as Teflon, with a $\gamma_A{}^{LW}$ of 17.9 mJ/m$^2$ (see Table 2.2) while water has a $\gamma_W{}^{LW}$ of 21.8 mJ/m$^2$ (see Table 2.1) its $\gamma_{AW}$ would be: $(\sqrt{17.9} - \sqrt{21.8})^2 + 2 \times 25.5 = 0.192 + 51 \approx 51.2$ mJ/m$^2$, which is quite close to the $\gamma_W{}^{AB}$ value for water (see Table 2.1).

With octane, with a $\gamma_A{}^{LW}$ value of 21.6 mJ/m$^2$ (see Table 2.1), which is extremely close to $\gamma_W{}^{LW} = -21.8$ mJ/m$^2$, the $\gamma_{AW}{}^{LW}$ part is equal to 0.0005 mJ/m$^2$, which may be neglected. This leaves $\gamma_{AW} = 2 \times 25.5 = 51$ mJ/m$^2$, which is exactly equal to the $\gamma_W{}^{AB}$ value for water, at 20 °C (see Table 2.1).

These two examples illustrate that the interfacial tensions between *apolar* materials and water tend to be around 51 mJ/m$^2$, as can already be seen in Girifalco and Good (1957), whilst from a paper by Fowkes (1963) the fact emerged (in an indirect manner) that 51 mJ/m$^2$ is also the value of $\gamma_W{}^{AB}$, or more accurately, in Fowkes' approach, to the value of $\gamma_W - \gamma_W{}^{LW}$, i.e., $72.8 - 21.8 = 51$ mJ/m$^2$ (for water at 20 °C). For a more precise explanation, see Eq. (4.3), where it can be seen that the only polar term [i.e., the far right one: $2\sqrt{(\gamma_W{}^+\gamma_W{}^-)}$] is indeed exactly the same as the term for $\gamma^{AB}$ of water: $2\sqrt{(\gamma_W{}^+\gamma_W{}^-)}$ (cf. Eqs. (1.2) and (2.3)). [It should be pointed out that Fowkes (1963) did not use the term, $\gamma^{LW}$, but employed, instead, $\gamma^d$, where the superscript, d, referred to the dispersion, or van der Waals–London component of $\gamma$. It was only since Chaudhury's (1984) study of the Lifshitz's (1955) approach that it became clear that, on a macroscopic level, what is true for van der Waals–London (dispersion) interactions is equally valid for all three van der Waals forces combined, later alluded to as Lifshitz–van der Waals (LW) interactions (which term was first used by van Oss, Chaudhury and Good, 1987, 1988). In any case, Fowkes (1963) was correct in asserting that van der Waals–London (dispersion) interactions are by far the most important of the three (see Chapter 1).]

Returning to the observation that the right-hand term of the $\gamma_{AW}$ expression for the interfacial tension between completely apolar (i.e., hydrophobic) materials or compounds, and water [i.e., $2\sqrt{(\gamma_w{}^+\gamma_w{}^-)}$], is identical with the expression for $\gamma_W{}^{AB}$, this is not a gratuitous coincidence, because in both cases the same term describes the AB attraction between the electron-acceptor and the electron-donor of water. When one multiplies the polar, right-hand term of Eq. (4.3) by a factor, $-2$, one obtains the part of $\Delta G_{iwi}{}^{AB}$, which for completely apolar molecules or particles, immersed in water, equals $-102$ mJ/m$^2$ and is represented by term 3 of the total $\Delta G_{iwi}$ equation treated in Chapter 5, Eq. (5.1). Term 3 of Eq. (5.1) stands for the hydrogen-bonding (AB) free energy of cohesion between water molecules, at 20 °C, which is equated with the hydrophobic energy of attraction between completely apolar (hydrophobic) molecules of particles, when immersed in water (see Chapter 5, Sub-sections 1.1–1.3). This applies therefore to the completely apolar Teflon particles as well as to the equally apolar octane molecules, when immersed in water. Similarly, the polar (AB) component, $\gamma^{AB}$, of the surface tension of water, is also equal to $2\sqrt{(\gamma_w{}^+\gamma_w{}^-)}$ so that, according to $\Delta G_{ww}{}^{AB} - -2\gamma_w{}^{AB}$, the latter is also equal to $-102$ mJ/m$^2$, being the AB, or the hydrogen-bonding free energy of cohesion between two water molecules at 20 °C.

## 2.2 Measurement of $\gamma_{AW}$ between *apolar* organic liquids and water

*Apolar* organic liquids are the only organic liquids whose (molecular) interfacial tensions with water still are correctly measurable via drop-shape or drop-weight methods. This methodology [see, e.g., Adamson (1990) and Hiemenz and Rajagopalam (1997)] was popular during most of the 20th century; see also Harkins (1952) as well as Girifalco and Good (1957) for lists of $\gamma_{iw}$ values for "oil–water systems" obtained with its use. However, upon further scrutiny by van Oss and Good (1996) and by van Oss, Giese and Good (2002) the $\gamma_{iw}$ values obtained by drop-shape or drop-weight methods *were found to be much too low for all polar organic liquids*; see Sub-section 3.1, below.

## 3. The Interfacial Tension between *Polar* Compounds or Materials and Water

### 3.1 Expression of $\gamma_{iw}$ between *monopolar* compounds and water

There is an entire category of polar compounds which are not dipolar, but rather *monopolar*, that is to say, such compounds tend to be strong electron-donors ($\gamma^-$) but have zero, or close-to-zero electron-accepticity ($\gamma^+$) (van Oss, Chaudhury and Good, 1987; van Oss, Giese and Wu, 1997). One of these monopolar electron-donating compounds is polyethylene oxide (PEO), also called polyethylene glycol (PEG), see Table 2.2. PEO is a monopolar electron-donor in the dry as well as in the dissolved state. Other such compounds are monopolar electron-donors when they are in the dried state but become dipolar when dissolved in water, with a small but usually non-negligible electron-accepticity. These comprise blood serum proteins such as human serum albumin and human serum immunoglobulin-G (IgG) (see Table 2.3), and carbohydrates such as sucrose, glucose, or the polymer, dextran (see Table 2.4). Monopolar electron-donors are often found in Nature, whilst monopolar *electron-acceptors* seldom appear in a free form. In all cases of monopolar Lewis bases, Eq. (4.2) is reduced to:

$$\gamma_{iw}^{\text{monopolar electron-donor}} = (\gamma_i^{LW} - \gamma_w^{LW})^2 + \sqrt{(\gamma_w^+\gamma_w^-)} - 2\sqrt{(\gamma_i^-\gamma_w^+)}. \quad (4.4)$$

There is no polar cohesion term in Eq. (4.4), because there is no $\gamma_i^+$ with which $\gamma_i^-$ can combine and there can be no adhesive term between a non-existent electron-acceptor of $\gamma_i$ and the electron-donor of water ($\gamma_w^-$). An important consequence of electron-donor monopolarity is that their $\gamma_i^{AB}$ is zero on account of the absence of $\gamma_i^+$.

Monopolar molecules such as PEO, various proteins, as well as monopolar particles, such as some clays (e.g., hectorite) and other mineral particles such as silica, glass and Hematite (cf. Table 2.6) are all exceedingly polar (and thus hydrophilic), as is shown by their positive $\Delta G_{SWS}$ values; see Chapter 5, Sub-section 1.1.

### 3.2 Measurement of $\gamma_{iw}$ between polar organic liquids and water

The measurement of interfacial tensions between water-immiscible liquids and water by means of drop-shape or drop-weight determinations was a popular approach for determining "oil–water" interfacial tensions during most of the 20th century.

For the methodology of this approach see, e.g., Adamson (1990) and Hiemenz and Rajagopalan (1997) and for lists of "oil–water" systems' $\gamma_{iw}$ values thus obtained, see Harkins (1952) and Girifalco and Good (1957).

As mentioned in Sub-section 2.2, above, for *apolar* aliphatic hydrocarbons (i.e., alkanes) the $\gamma_{iw}$ values for "oil–water" systems listed by Girifalco and Good (1957) are still perfectly valid. However, as noted by van Oss and Good (1996) and van Oss, Giese and Good (2002) the $\gamma_{iw}$ values obtained with *polar organic liquids* via the drop-shape or drop-weight approach are much too low and should not be used.

Van Oss and Good (1996) demonstrated that for all organic compounds, the interfacial tension with water, $\gamma_{iw}$, is related to their aqueous solubility (s, expressed in mol fractions) as well as to the contactable surface area ($S_c$) between two molecules of the compound when immersed in water:

$$-2\gamma_{iw} \cdot S_c = kT \cdot \ln s, \tag{4.5}$$

where k is Boltzmann's constant ($k = 1.38 \times 10^{-23}$ J/ K), while T is the absolute temperature in Kelvin. Furthermore, $S_c$ is the contactable surface area between two molecules, i, when immersed in water and the solubility, s, should be expressed in dimensionless "mol fractions." Using Eq. (4.5), it could be demonstrated that for *alkanes* the $\gamma_{iw}$ values measured by drop-shape or drop-weight methods (see the preceding Sub-section 2.2) were entirely in accord with those found via the alkanes' aqueous solubilities. However, for *polar* organic compounds (including benzene) the $\gamma_{iw}$ values obtained via drop-shape or drop-weight methods were invariably too low, as compared with the molecular $\gamma_{iw}$ values derived from their aqueous solubilities (van Oss, Wu et al., 2001; van Oss, Giese and Good, 2002). By comparing the solubilities found for alkyl alcohols with the solubilities calculated from the drop-shape derived $\gamma_{iw}$ values it was found that, e.g., for *n*-octyl alcohol, the calculated aqueous solubility would have been *914* times greater than its real experimental solubility (van Oss and Good, 1996). This large discrepancy indicated that for *polar* organic compounds the drop-shape or drop-weight methods for the derivation of *interfacial tensions with water* leads to grossly erroneous results and should not be used for that purpose. On a *very macroscopic scale* however, the drop-shape or drop-weight methodology will give accurate results for the measurement of interfacial tensions between *whole drops of the polar liquid and water*, in case there exists a need for such data. The reason for the non-applicability of drop-shape or drop-weight measurements with an aim at obtaining interfacial tensions between *polar* organic liquids and water is treated in the following sub-section.

## 3.3 The zero time dynamic interfacial tension between polar organic liquids and water

When using drop-shape or drop-weight methods with polar organic compounds, from the very moment such polar liquids are immersed in water only those molecules that actually dissolve in the water remain suspended as single, separate molecules which have their polar as well as their apolar moieties freely exposed to the water on all sides, via their water-interfaces. The aggregate interfacial tensions with water of the polar plus the apolar moieties of such single polar molecules thus contribute to the total aqueous solubility of each complete polar molecule as long as they remain in solution as separate entities. When an excess of polar molecules is immersed in water (which is often the case with sparsely soluble organic compounds), the non-dissolved molecules remain agglomerated as whole drops, suspended in the water. Now, these residual drops unavoidably and exceedingly quickly undergo hysteresis caused by the virtually instantaneous orientation of their polar moieties toward the drops' interface with water (van Oss, Giese and Good, 2002). Thus, at the very moment of their immersion in water drops of a polar organic liquid will, at time zero, still show the correct molecular value for their interfacial tension with water, but within picoseconds (Luzar and Chandler, 1996) hydrogen bonds between water molecules and the polar moieties of the polar organic compounds will form (and break, and reform again), leaving the apolar moieties of the organic molecules more and more inside the drop, away from the water interface. Thus, in an extremely short time-lapse more and more of the polar moieties of the polar organic molecules of each *drop* will have migrated toward its water interface, which very quickly lowers its measurable interfacial tension with water to macroscopic-scale values that are much lower than the molecular, microscopic-scale values occurring between water and single dissolved polar organic molecules. This is because with single polar *molecules* dissolved in water, their total interfacial tension with water is *averaged* between the polar and the apolar moieties of the solute molecule, whilst with *whole drops* of such polar liquids, the interface on the water side almost exclusively "sees" only the polar moieties of such molecules within barely picoseconds after the drops' immersion. For that reason the microscopic-scale interfacial tension between singly dissolved molecules of polar organic liquids and water have been alluded to as "zero time dynamic" interfacial tensions with water, or $\gamma_{iw}{}^{0}$ (van Oss, Giese and Good, 2002). Clearly, $\gamma_{iw}{}^{0}$ values of polar organic liquids cannot be accurately measured by drop-shape or drop-weight methods, because it is not feasible to measure such $\gamma_{iw}{}^{0}$ values via drop-shape or drop-weight observations within the first one or two picoseconds before hysteresis sets in, immediately after a polar organic drop is immersed in water.

Thus, other methods must be applied for measuring the zero time dynamic, molecular-scale, interfacial tension between polar organic molecules and water, i.e., $\gamma_{iw}{}^{0}$.

## 3.4 Determination of $\gamma_{iw}{}^{0}$ via the aqueous solubility of i

Adopting the aqueous solubility approach indicated in Sub-section 3.2 and using Eq. (4.5) given in that sub-section, the correct molecular interfacial tensions with water were established for: benzene, chloroform, ethyl ether and ethyl-acetate (see Table 2.1), as well as for sucrose, glucose and dextran (in the dissolved state in water; see Table 2.4; see also Docoslis et al., 2000). Also determined were the $\gamma_{iw}{}^{0}$ values for toluene and xylene (van Oss, Giese and Good, 2002). It should be noted that in Tables 2.1–2.6, all $\gamma_{iw}{}^{0}$ values can be obtained by multiplying $\Delta G_{SWS}$ by a factor −0.5.

It should be reiterated that it is only with *apolar* organic liquids that one will find that $\gamma_{iw}{}^{0}$ determined via their aqueous solubility or measured via drop-shape or drop-weight approaches gives identical results. With *polar* organic liquids drop-shape or drop-weight methods should not be used (van Oss and Good, 1996).

## 3.5 Derivation of $\gamma_{iw}{}^{0}$ from the polar equations for $\gamma_{iw}$, after having determined the components and parameters of $\gamma_i$

One can obtain the value for $\gamma_{iw}{}^{0}$ by using the polar equation for $\gamma_{iw}$ (Eq. (4.2)). This requires that one knows, or determines, the values for $\gamma_i^{LW}$, $\gamma_i^{+}$ and $\gamma_i^{-}$ obtainable via contact angle measurements and the polar version of the Young–Dupré equation (see Eq. (2.15B) and Chapter 2, Sub-section 1.6.3 for direct contact angle measurement or Chapter 2, Sub-section 1.6.4 for wicking). This also requires the necessary information on the $\gamma_L^{LW}$, $\gamma_L^{+}$ and $\gamma_L^{-}$ values for all three contact angle liquids used in the contact angle measurements (e.g., for diiodomethane, water and glycerol; see Table 2.1); see the following sub-section.

## 3.6 Determination of $\gamma_{iw}$ via the Young equation, the polar properties of $\gamma_i$ and the water contact angle measured on material, i

In some cases it may be simpler to determine $\gamma_{iw}$ using the Young equation (cf. Eq. (2.14)):

$$\gamma_w \cdot \cos\theta = \gamma_{iw} + \gamma_i. \tag{4.6}$$

This only requires one contact angle determination with drops of water (to obtain $\cos\theta$), plus knowledge of $\gamma_w$ (which is 72.8 mJ/m$^2$ at 20 °C) and knowledge of the polar properties of material, i, as outlined in the preceding Sub-section 3.5, above, while avoiding use of the more elaborate polar $\gamma_{iw}$ equation (Eq. (4.2)).

CHAPTER FIVE

# The Interfacial Tension/Free Energy of Interaction Between Water and Identical Condensed-Phase Entities, i, Immersed in Water, w

## Contents



## 1. The $\Delta G_{iwi}$ Equation Pertaining to Identical Entities, i, Immersed in Water, w

The apolar + polar version of the (interfacial) $\Delta G_{iwi}{}^{IF}$ equation, comprising the Lifshitz–van der Waals (LW), plus the Lewis acid–base (AB) interactions follows

*Interface Science and Technology*, Volume 16
ISSN 1573-4285, doi: 10.1016/S1573-4285(08)00205-6

$$\Delta G_{iwi}^{IF} = -2\gamma_{iw} = -2(\underset{1}{\sqrt{\gamma_i^{LW}} - \sqrt{\gamma_w^{LW}})^2} - \underset{2}{4\sqrt{(\gamma_i^+\gamma_i^-)}} - \underset{3}{4\sqrt{(\gamma_w^+\gamma_w^-)}}$$

$$+ \underset{4}{4\sqrt{(\gamma_i^+\gamma_w^-)}} + \underset{5}{4\sqrt{(\gamma_i^-\gamma_w^+)}}, \tag{5.1}$$

where $\Delta G_{iwi}^{IF} = \Delta G_{iwi}^{LW} + \Delta G_{iwi}^{AB}$, but does not include $\Delta G_{iwi}^{EL}$ (see also Eq. (2.11)).

## 1.1 $\Delta G_{iwi}^{LW}$, or the apolar component of $\Delta G_{iwi}^{IF}$

Term 1 of Eq. (5.1) denotes the apolar (LW) component of the interaction between two molecules or surfaces of compound, i, immersed in water. This term is always attractive (i.e., the value of $\Delta G_{iwi}^{LW}$ is negative), or zero, see also Eq. (2.11). As can be seen in Table 2.1, various alkanes, which are completely apolar, also have rather low $\gamma_i^{LW}$ values, which are in the neighborhood of the $\gamma_w^{LW}$ for water ($\gamma_w^{LW}$ = 21.8 mJ/m$^2$, at 20 °C), so that term 1 of Eq. (5.1) then is quite small, or even zero (e.g., in the case of octane).

## 1.2 $\Delta G_{iwi}^{AB}$, or the polar component of $\Delta G_{iwi}^{IF}$

It is important to realize that even when *completely non-polar* entities are immersed in water, *polar* (AB) interactions caused by water alone remain nonetheless of crucial importance. In such cases term 3 is of particular significance; see also Eq. (4.3). Term 3 denotes the polar energy of cohesion of water, which is *always* active in aqueous media and also happens to be the driving force of the ("hydrophobic") attraction between apolar entities immersed in water, i.e., the *hydrophobic effect*.

It should be noted that AB forces in water are operative for *dipolar*, *monopolar* as well as for *apolar* compounds or materials. Referring to the polar part of Eq. (5.1), with *dipolar* materials *both* terms 2 and 3 are *attractive*; they indicate, respectively, the polar free energy of cohesion of compound or material, i (term 2) and the polar free energy of cohesion of water (term 3). For *monopolar* materials however, term 2 is zero but term 3 remains negative (i.e., attractive). It should be stressed that for *apolar* materials, i, term 2 may be negative (attractive), or zero, but term 3, expressing the polar cohesion of water is always negative (attractive); it never becomes zero and is always quite large. At 20 °C, term 3 for water amounts to −102 mJ/m$^2$ and is permanently active, at that strength, for apolar as well as for monopolar and dipolar materials.

For *dipolar* materials terms 4 and 5 are both repulsive (i.e., positive); they indicate the polar adhesion of i to water, in other words they are the two hydration terms for i. Term 5 is usually much larger than term 4, because most polar compounds or materials, i, are preponderantly Lewis bases (i.e., electron-donors), with a much smaller, or even zero Lewis accepticity, in which latter case i is *monopolar* (van Oss, Giese and Wu, 1997); see Chapter 4, Sub-section 3.1.

## 1.3 The use of $\Delta G_{iwi}^{IF}$ in the quantitative definition of hydrophobicity and hydrophilicity

In this sub-section on the quantitative identification of hydrophobicity and hydrophilicity, two semi-quantitative indicators of these conditions are treated first, in the guise of the $\gamma_i^-$ value (Sub-section 1.3.1) and the value of the free energy of hydration of i, i.e., $\Delta G_{iw}$ (Sub-section 1.3.2). Treatment of the only rigidly quantitatively precise expression, in the form of $\Delta G_{iwi}^{IF}$ (see Eq. (5.1)) is given in Sub-section 1.3.3, below.

### 1.3.1 The $\gamma_i^-$ value as an indicator of the hydrophobicity or hydrophilicity of condensed-phase material (i) hydrophilicity of i

The $\gamma_i^-$ value of condensed-phase materials (i) can be a convenient but qualitative indicator of their hydrophobicity or hydrophilicity. Strongly hydrophobic (apolar) materials tend to have very low (or zero) $\gamma_i^-$ values. (The $\gamma_i^+$ values of virtually all organic, biological and even most mineral compounds or materials, usually are zero, or very low, whether they be hydrophobic or hydrophilic.) For the $\gamma i^+ = 0$ values linked to apolar liquids, see Table 2.1 for the alkanes, and Table 2.2 as far as a number of completely apolar polymers is concerned and see Table 2.1, 2.2, 2.3 and 2.4 which contain some monopolar electron-donating materials or compounds and Tables 2.5 and 2.6 containing some close-to-monopolar materials with very low $\gamma_i^+$ but with sizable $\gamma_i^-$ values.

The cut-off between hydrophobic and hydrophilic compounds or materials with a $\gamma_i^{LW} \approx 40$ mJ/m$^2$ is about at $\gamma_i^- \approx 28.3$ mJ/m$^2$; see for instance polyvinyl pyrrolidone, with $\gamma_i^- = 29.7$ mJ/m$^2$, which makes it just barely hydrophilic; see Table 2.2 and see also the smectite clay, Swy-1 (Table 2.5) which is almost, but not quite hydrophilic with $\gamma_i^- = 29.2$ mJ/m$^2$, because it has also a small but non-negligible $\gamma_i^+$ of 1.5 mJ/m$^2$.

Very hydrophilic compounds or materials have quite elevated $\gamma_i^-$ values; see, e.g., glycerol and formamide (Table 2.1); the proteins, hydrated human serum albumin and IgG, as well as bovine fibrinogen (Table 2.3) and the exceedingly hydrophilic polymer, polyethylene oxide (PEO) (Table 2.2). See also the carbohydrates, sucrose, glucose and the latter's linear polymer, dextran, shown in Table 2.4, all with high $\gamma_i^-$ values and accompanied by zero $\gamma_i^+$ values in the dry state, but manifesting non-zero $\gamma_i^+$ values when in aqueous solution; see Docoslis et al. (2000). Finally, see the high $\gamma_i^-$ values for the clays, bentolite and kaolinite, as well as for mica and silica (Table 2.5) and for glass and hematite (Table 2.6), which are also quite hydrophilic.

However, due to small but non-negligible variations in the $\gamma_i^{LW}$ values among different compounds or materials, which also contribute to the total $\Delta G_{iwi}^{IF}$ value, the estimation of the degree of hydrophobicity/hydrophilicity of compound or material, i, judging by the observation that their value of $\gamma_i^-$ is below or above approximately 28 to 29 mJ/m$^2$, tends to be less than precise. It is only fairly reliable when $\gamma_i^-$ is between zero and about 20 mJ/m$^2$, when one may conclude with some confidence that i is hydrophobic. Finally, when $\gamma_i^-$ is significantly greater than, say, 30 mJ/m$^2$ one may reasonably safely conclude that i is most likely hydrophilic. Nonetheless, even within these limitations, the value of $\gamma_i^-$ can never be

**Table 5.1** $\Delta G_{iwi}^{IF}$, $\gamma_i^-$ and $\Delta G_{iw}$ values for hexane, agarose, polyvinyl pyrrolidone and polyethylene oxide[a], as well as for the water–air interface[b], in mJ/m², at 20 °C

| Compound | | $\Delta G_{iwi}^{IF}$ | $\gamma_i^{-c}$ | $\Delta G_{iw}$ |
|---|---|---|---|---|
| The water–air interface | Hydrophobic | −145.6 | 0 | 0 |
| Hexane | Hydrophobic | −102.4 | 0 | −40.0 |
| Agarose | Hydrophobic | −3.2 | 26.9 | −112.2 |
| Transition point between hydrophobicity and hydrophilicity | | 0 | ≈28.3 | ≈ −113.0 |
| Polyvinyl pyrrolidone | Hydrophilic | +1.0 | 29.7 | −116.6 |
| Polyethylene oxide | Hydrophilic | +52.5 | 64.0 | −142.0 |

*a* Data from van Oss (1994, 2006).

*b* See Chapter 9.

*c* In all cases, $\gamma_i^+ = 0$.

a genuine quantitative measure for the *degree* of hydrophobicity or hydrophilicity of compound or material, i, although it can often serve as a useful rough estimate of the hydrophobicity or hydrophilicity of i (cf. Table 5.1).

### 1.3.2 The free energy of hydration of i, $\Delta G_{iw}$, as an indicator of the hydrophobicity or hydrophilicity of i

Table 5.1 also shows the behavior of the free energy of hydration of i, $\Delta G_{iw}$, from the hyper-hydrophobic water–air interface (van Oss et al., 2005), to one of the most hydrophilic compounds known, i.e., polyethylene oxide (PEO) [also often called polyethylene glycol (PEG)]. The $\Delta G_{iw}$ values in Table 5.1 vary from $\Delta G_{iw} = 0$ (for the paradoxically totally hydrophobic and non-hydrated water–air interface) to the strongly hydrated PEO surface, with a $\Delta G_{iw}$ value of −142 mJ/m². The transition between hydrophobicity and hydrophilicity lies at around $\Delta G_{iw} = -113$ mJ/m2 (see Table 5.1). Like $\gamma_i^-$, the value for $\Delta G_{iw}$ furnishes a mildly useful but still only semi-quantitative estimate of the degree of hydrophobicity/hydrophilicity of material or compound, i.

### 1.3.3 The interfacial free energy of interaction between two surfaces or molecules, i, immersed in water, $\Delta G_{iwi}^{IF}$, represents the surface-thermocynamic hydrophobicity/hydrophilicity scale

$\Delta G_{iwi}^{IF}$ represents the surface-thermodynamic definition of the hydrophobicity/hydrophilicity of condensed-phase materials. A *negative* $\Delta G_{iwi}^{IF}$ not only indicates that i is hydrophobic but also gives the quantitative value of its hydrophobicity. This is because two molecules or particles, i, attract one another when immersed in water with precisely the free energy of the negative value of $\Delta G_{iwi}^{IF}$, which in its turn correlates directly with the aqueous solubility of *dissolved molecules* (see Eq. (5.2), below, as well as Chapter 7) or with the degree of diminished stability of *suspended particles* (see Chapter 8).

A *positive* $\Delta G_{iwi}^{IF}$ value on the other hand indicates that i is hydrophilic. This is because when $\Delta G_{iwi}^{IF} > 0$, two molecules or particles, i, repel one another and the exact positive value for $\Delta G_{iwi}^{IF}$ indicates the precise amount of free energy of repulsion between two molecules or particles, i, when immersed in water. For immersed *molecules any positive value* of $\Delta G_{iwi}^{IF}$ is tantamount to complete aqueous solubility (see Chapter 7), whilst for immersed *particles* the precise positive value of $\Delta G_{iwi}^{IF}$ is a quantitative indication of the degree of aqueous suspension stability of i (see Chapter 8).

The complete aqueous *solubility of molecules* of i, when $\Delta G_{iwi}^{IF} > 0$ may be explained as follows: Even when $\Delta G_{iwi}^{IF} = 0$, according to the solubility equation:

$$\Delta G_{iwi}^{IF} \cdot S_c = kT \cdot \ln s \tag{5.2}$$

(cf. Eq. (4.5)), the aqueous solubility (s) of compound (i) is equal to 1 mol fraction which, in water, would mean that the aqueous solubility of compound, i, s = 55.6 mol/L which is a feat that is hard to achieve under conditions of ambient pressure and temperature for any solute consisting of molecules that are bigger than those of water (van Oss and Giese, 2004; van Oss, 2006). Thus, once $\Delta G_{iwi}^{IF} \geqslant 0$, one has ample aqueous solubility of i. In the case of PEO-6000 for example, that solubility amounts to s = 15.6 mol fractions, indicating that for each $H_2O$ molecule there could be 15.6 PEO molecules dissolved in the water, signifying that the aqueous solubility of PEO is quasi-infinite and in practice only limited by phenomena such as viscosity or, at increased temperatures, by the influence of the $\theta$-point of PEO (which can be as low as 50 °C at which point the secondary structure of the PEO itself changes by exposing more of its polyethylene backbone, which leads to insolubility) see van Oss and Giese (2004); van Oss (2006).

The aqueous solubilities of the many compounds and materials that have a positive $\Delta G_{iwi}^{IF}$ (see Tables 2.1 to 2.4), lack the real quantitative meaning of the word which it does have when $\Delta G_{iwi} < 0$. As far as compounds with $\Delta G_{iwi}^{IF} > 0$ are concerned, one simply describes their solubility as: "completely soluble in water."

For *particles* suspended in water a positive $\Delta G_{iwi}^{IF}$ value (see Tables 2.5 and 2.6) has a more precise meaning in that such a positive $\Delta G_{iwi}^{IF}$ then denotes the *degree* of *stability* of their aqueous suspension, e.g., by means of XDLVO analysis with energy vs distance plots; see Chapter 3. Or, as a rough estimation, one may assume that when $\Delta G_{iwi}^{IF}$ (at "contact") attains a value of about +10 kT, one has "an energy barrier which prevents the particles from colliding" (Overbeek, 1977), see also Chapter 8.

Please note: The $\Delta G_{iwi}^{IF}$ values of all materials and compounds given in Tables 2.1 to 2.6, are listed in the far right-hand columns.

## 2. Mechanism of Hydrophobic Attraction in Water

### 2.1 $\Delta G_{iwi}^{IF}$ and the hydrophobic effect

Returning to the $\Delta G_{iwi}^{IF}$ equation (Eq. (5.1)) and taking octane, immersed in water, as an example, where $\gamma_i^{LW}$ for octane amounts to 21.6 mJ/m$^2$, whilst for water $\gamma_w^{LW}$ equals 21.8 mJ/m$^2$, it is easily seen that term 1 equals only 0.0005 mJ/m$^2$, which may be taken as zero. Then, as octane is completely apolar, its $\gamma_i^+$ and $\gamma_i^-$ are zero, which reduces terms 2, 4 and 5 to zero, leaving only term 3, which for water is always present and which at 20 °C amounts to exactly $-102$ mJ/m$^2$. Term 3 represents the free energy of Lewis acid–base (in this case hydrogen-bonding) free energy of cohesion of water. Thus term 3, being the only non-zero term left for $\Delta G_{iwi}^{AB}$ when denoting the free energy of interaction between completely apolar molecules (such as octane), when immersed in water, is the driving force for the hydrophobic attraction between apolar molecules or particles, in water, i.e., for the "hydrophobic effect." The negative value of term 3, at $-102$ mJ/m$^2$, at 20 °C decreases gradually with an increase in temperature, to about $-85.4$ mJ/m$^2$ at 100 °C (van Oss, 1994, 2006). It should be reemphasized that term 3 of the $\Delta G_{iwi}^{IF}$ equation, being an intrinsic property of water itself, is always there, and always attractive. Thus the hydrophobic attraction is always present, for all interactions occurring in water. In some cases of polar compounds, immersed in water, the total $\Delta G_{iwi}^{IF}$ value can even be somewhat more negative than $-102$ mJ/m$^2$. This happens when a polar compound, i, has a $\gamma_i^{LW}$ value which is significantly higher than the 21.8 mJ/m$^2$ for water; see, e.g., the case of diiodomethane (Table 2.1), where $\Delta G_{LWL}^{IF} = -112$ mJ/m$^2$. Even with strongly polar, hydrophilic compounds or particles, immersed in water, their net hydrophilicity can only manifest itself when their free energy of hydration (mainly incorporated in the positive term, 5) is large enough to overcome the always present $-102$ mJ/m$^2$ of term 3; see Section 3, below.

The propagation at a distance of the hydrophobic attraction in water has been treated in Chapter 3, Sub-section 3.4; see also Chapter 10, Section 2.

## 2.2 Increasing or decreasing the hydrophobizing capacity of water

### 2.2.1 Increasing the hydrophobizing capacity of water

In Sub-section 1.3.3, above, the degree of hydrophobicity and hydrophilicity of material or compound, i, when immersed in water, have been defined as the quantitative values of $\Delta G_{iwi}^{IF}$, pertaining to hydrophobicity of i when negative (i.e., attractive) and to the hydrophilicity of i when positive (i.e., repulsive). As to hydrophobicity, it was determined in preceding Sub-section 2.1 (see above) that the driving force for hydrophobic attraction (i.e., the "hydrophobic effect") is term 3 of Eq. (5.1), which stands for the hydrogen-bonding (Lewis acid–base) free energy of cohesion of water, amounting to $-102$ mJ/m$^2$ at 20 °C, which is therefore, by its very nature, always there. There are two different approaches to effect an increase in the AB-driven free energy of cohesion of water: (I) By increasing the polar free energy of cohesion of water through cooling, or: (II) By adding soluble polar solutes to the water, which have an AB free energy of cohesion that is even higher than that of water.

***Ad* I:** A decrease in temperature by, say, 20 °C would only increase the total free energy of cohesion of water by roughly 2% (see van Oss, 1994, p. 300; van Oss, 2006, p. 95). The influence on the AB free energy of cohesion of water wielded by *cooling* is therefore minimal.

*Ad* II: Increasing the AB free energy of cohesion of water, and thus its hydrophobizing capacity, by the admixture of water-soluble, strongly self-binding solutes can be considerable. There are two classes of solutes that can raise the AB free energy of cohesion of water, i.e., sugars and neutral salts.

The admixture to water of a sugar such as sucrose at a concentration of 55% (wt%) increases the free energy of cohesion of water by about 6.4 $mJ/m^2$ (Docoslis et al., 2000; see also the Handbook of Chemistry and Physics, CRC Press, 51st ed., 1970/71, p. F-28). Although that increase is not enormous, Docoslis et al. (2000) showed that the addition of 55% sucrose to originally stable aqueous suspensions of kaolinite and silica particles gave rise to the destabilization (flocculation) of these suspensions. However, admixture of 55% sucrose to an aqueous immunoglobulin G (IgG) solution did not cause the IgG molecules to precipitate (Docoslis et al., 2000, unpublished observation).

The addition of approximately 25% (wt%) of neutral salts can increase the (AB + EL) free energy of cohesion of water by anywhere from about 5% (with LiCl) to up to 26% (with NaCl), which is therefore very much dependent upon the nature of the neutral salt used (see Handbook of Chemistry and Physics, 1970/71, *vide supra*). It is therefore quite understandable that the addition of fairly high concentrations of neutral salts is more effective than the addition of high concentrations of sugars in precipitating proteins from their aqueous solutions (where in the case of added salts this procedure is called "salting out"). In addition, it should be noted that contrary to sugar solutions, concentrated salt solutions in water also tend to reduce any electrokinetic ($\zeta$) potentials suspended particles or dissolved proteins would normally have when immersed in aqueous media at low ionic strengths (see Chapter 3, Sections 5–7, and Eqs. (3.5) and (3.6)). High salt concentrations not only reduce the electrostatic repulsion between suspended particles or dissolved protein molecules but it also concomitantly reduces the hydrophilicity of these particles or proteins (by reducing the values of their $\gamma^-$ parameters; see Chapter 3, Sub-section 6.3 and Section 7). These are additional reasons why high salt concentrations are more effective in increasing the hydrophobizing capability of water than high sugar concentrations. Nonetheless, both high salt and high sugar concentrations have been used for centuries for conserving foods in aqueous solutions or suspensions, where the strongly hydrophobizing conditions inhibit, *inter alia*, most enzymatic reactions as well as the growth of most common bacteria.

### 2.2.2 Decreasing the hydrophobizing capacity of water

To begin with, like the decrease in temperature treated in Sub-section 2.2.1, above, an *increase* in temperature also has only a slight influence on term 3 of Eq. (5.1). An increase in temperature of, say, 20 °C, will only decrease the value of term 3 to about $-98$ $mJ/m^2$, i.e., by only approximately 4%. This is in contrast with the considerable influence on the (AB) free energy of adhesion (hydration) when one increases the temperature of solutions of *hydrophilic* solutes (see Sub-section 3.2.1, below). To decrease the hydrophobizing capability of water one must decrease its polar (AB) free energy of cohesion, i.e., one must decrease the negative value of term 3 of Eq. (5.1). To that effect it should first be noted that term 3 is equal to

$2\gamma_w^{AB}$ (see Eq. (1.2)). In other words, one should add something to the water that reduces the value of its $\gamma_w^{AB}$, which normally amounts to 51 mJ/m$^2$ at 20 °C. A few candidates for this are water-miscible solvents such as methanol, ethanol, ethylene glycol or acetonitrile, all with a much lower $\gamma_L^{AB}$ than the 51 mJ/m$^2$ of water. See also Chapter 6, Sub-sections 2.1 and 2.2, for chromatographic applications of methods to decrease (or increase) the hydrophobizing capability of water.

# 3. Mechanism of Hydrophilic Repulsion in Water

## 3.1 $\Delta G_{iwi}^{IF}$ and hydrophilic repulsion, or "hydration pressure"

To achieve a net repulsion in water, one must first overcome the always-present, built-in hydrophobic attraction of water, due to its AB (hydrogen-bonding) free energy of cohesion of −102 mJ/m$^2$, as described by term 3 of Eq. (5.1). A net mutual repulsion in water can only be achieved by strongly hydrophilic cells, particles or molecules, which also tend to be monopolar, or close-to-monopolar electron-donors (i.e., Lewis bases); see van Oss, Giese and Wu (1997). Thus, of the two hydration terms of Eq. (5.1), i.e., terms 4 and 5, practically speaking only term 5 is relevant, because term 4 depends on the influence of $\gamma_i^+$ which in the vast majority of cases is zero, or very small.

Using polyethylene oxide (PEO) as an example (see Table 2.2), with a $\gamma_i^{LW}$ of 43 mJ/m$^2$, one has a $\Delta G_{iwi}^{LW}$ of −7.1 mJ/m$^2$ (i.e., term 1 of Eq. (5.1)). One also has the cohesive term for water in the guise of term 3, worth −102 mJ/m$^2$. Thus, in total, for any repulsion to be possible between two PEO molecules, immersed in water, hydration term 5 has to be greater than $7.1 + 102 = +109.1$ mJ/m$^2$. Now, with a $\gamma_i^+ = 0$ and a $\gamma_i^-$ of 64 mJ/m$^2$, PEO's term 5 amounts to $+4\sqrt{(\gamma_i^+\gamma_i^-)} = 4\sqrt{(64 \times 25.5)} = +161.6$ mJ/m$^2$ of hydration energy. Then the net free energy of interaction between two PEO molecules when immersed in water, $\Delta G_{iwi}^{IF} = 161.6 - 109.1 = +52.5$ mJ/m$^2$ (see also Table 2.2).

## 3.2 Increasing or decreasing the hydrophilic repulsion occurring in water

### 3.2.1 Increasing the hydrophilic repulsion occurring in water

To *increase* the hydrophilic repulsion which occurs in water one needs to augment the value of term 5 of Eq. (5.1). This is most effectively done by increasing the temperature of the aqueous medium, which increases the $\gamma_w^+/\gamma_w^-$ ratio ($=r_w$) of water. At 20 °C, $r_w$ is assumed to be equal to unity (see Chapter 2, Section 2). Then, heating the water to a temperature of 38 °C, the value of $r_w$ increases to 1.75 (van Oss, 1994, p. 301; van Oss, 2006, p. 94). Thus, upon heating, $\gamma_w^+$ has increased from 25.5 mJ/m$^2$ (at 20 °C) to 32.4 mJ/m$^2$ (at 38 °C), which increases the value of term 5 of Table 5.1 by 27%, if one may assume that the $\gamma_i^-$ value for PEO itself does not change significantly when increasing the temperature from 20 to 38 °C.

Thus an increase in temperature increases the positive value of $\Delta G_{iw}^{IF}$ for hydrophilic materials, or it decreases the negative value of $\Delta G_{iwi}^{IF}$ for mildly hydrophobic, partly polar materials. This is of special interest for compounds such as sugars (sucrose, glucose, etc.) which have a fairly high aqueous solubility (see, e.g., Stephen and Stephen, 1963), but which are not infinitely soluble in water, so that their $\Delta G_{iwi}^{IF}$ values remain somewhat on the negative side (Docoslis et al., 2000) and which therefore still classifies them in the "hydrophobic" category; see Sub-section 1.3.3, above; see also Chapters 7 and 9, below. An increase in temperature thus increases the aqueous solubility of non-infinitely soluble solutes such as simple sugars. It is also well known that an increase in temperature increases the ease of dissociation (due to an increase in term 5-associated repulsion; see Eq. (5.1), above), as for instance in washing various materials, even without soap, with warm rather than with cold water.

### 3.2.2 Decreasing the hydrophilic repulsion occurring in water

To *decrease* the hydrophilic repulsion occurring in water, cooling would be the most effective approach: By cooling the aqueous medium from 20 to 0 °C, the value of the $\gamma_w^+$ of water decreases from 25.5 to 19.0 mJ/m$^2$, i.e., by 25.5% (van Oss, 2006, p. 96).

The admixture of high concentrations of neutral salts can also help in decreasing the positive value of $\Delta G_{iw}^{IF}$, by increasing the negative value of term 3 (i.e., the hydrophobic effect term; see Sub-section 2.2.1, above). Although this approach does not involve term 5 of Eq. (5.1), it does play a role in the attachment of hyper-hydrophilic proteins (such as IgA) onto hydrophobic chromatography beads by indirectly lowering the positive value of $\Delta G_{iwi}$ *via* an increase in the negative value of term 3 through a salting-out effect; discussed in Chapter 6, Sub-section 2.4.2 on hydrophobic interaction chromatography.

## 4. Osmotic Pressures of Apolar Systems as well as of Polar Solutions, Treating Aqueous Solutions in Particular

Equation (5.1), above, plays a crucial role in expressing the osmotic pressure ($\Pi$) of polar polymers, when dissolved in water.

### 4.1 Osmotic pressure in apolar systems

The osmotic pressure, $\Pi$, caused by an *apolar* solute dissolved in an apolar (usually organic) solvent is completely described by van 't Hoff's equation:

$$\Pi = RTc \tag{5.3}$$

where R is the gas constant (R = 8.3143 J/degree/mole (SI), or in cgs: 8.3143 × 10$^7$ ergs/degree/mole; T is the absolute temperature in degrees Kelvin and c is the concentration in moles/L. At c = 1 molar (at 20 °C), $\Pi$ = 24.361 ×

$10^5$ Newton/m$^2$ = 2.4361 Mega Pascal (MPa) = 24.36 × $10^6$ dynes/cm$^2$ = 24.85 kg/cm$^2$ = 353.4 pounds per square inch (p.s.i.) ≈ 24.0 atmospheres [van Oss, 1994, Chapter V; 2006, Chapter VI]. In such an apolar liquid the osmotic pressure of 1 molar apolar solute is caused by 6.02 × $10^{23}$ dissolved molecules per liter solvent, resulting in the above-mentioned osmotic pressure of 2.4361 MPa. (6.02 × $10^{23}$ is Avogadro's number of molecules per mole.) Under these conditions one may state that the osmotic pressure of a one molar solution of apolar molecules in one liter of an apolar solvent is caused by the thermal bumps of 6.02 × $10^{23}$ molecules, where each molecule is endowed with a Brownian motion energy of 1 kT, and where k is Boltzmann's constant of 1.38 × $10^{-23}$ J/degree and T is the absolute temperature in degrees Kelvin. [It may be noted that $k = R/(6.02 \times 10^{23})$, so that $R$ = Avogadro's number times k.]

## 4.2 Osmotic pressure of aqueous polymer solutions

It should be recalled that apolar or only slightly polar polymers are *not soluble in water*, because they are *hydrophobic*, so that they *attract* each other with up to −102 mJ/m$^2$ when immersed in water; see term 3 of Eq. (5.1); see also Section 2, above. On the other hand, with *polar* polymers, which are *hydrophilic*, the positive value of term 5 of Eq. (5.1) is significantly greater than the negative value of term 3 of that equation, which causes them to have a positive $\Delta G_{iwi}{}^{AB}$ value, so that they repel one another in water, which favors their dissolution, see Chapter 7, Section 3. Thus, *in addition to obeying the van 't Hoff equation* (Eq. (5.3)) which takes their (repulsive) Brownian motion energy into account, *one also must include their polar free energy of mutual repulsion when expressing their total osmotic pressure*, see van Oss, 1994, Chapters V and XV; 2006, Chapters VI and XIX). In these chapters of 1994 and 2006, the mutual repulsion energies of dissolved hydrophilic polymers were calculated by using the Flory–Huggins theory of polymer solubility (Flory, 1953). This worked remarkably well when compared with experimentally measured osmotic pressures of various aqueous solutions of exceedingly hydrophilic polyethylene oxides (PEO) of different molecular weights (Arnold et al., 1988; see also van Oss, Arnold et al., 1990). It should also be mentioned that in van Oss, Arnold et al. (1990) the Flory–Huggins $\chi$-parameter (Flory, 1953) had been taken into account, after demonstrating that:

$$\chi_{iw} = -S_c \cdot \Delta G_{iwi}{}^{AB}, \tag{5.4}$$

where $\chi_{iw}$ is expressed in kT units and where $S_c$ is the contactable surface area between two molecules of water-soluble polymer, immersed in water (see Chapter 7, Section 1). The complete apolar + polar version of $\Delta G_{iwi}{}^{IF}$ is given in Eq. (5.1), above.

## 4.3 Osmotic pressure of linear polar polymers, dissolved in water

For a linear, hydrophilic, water-soluble polymer such as PEO, the osmotic pressure of its aqueous solution may be written as:

$$\Pi = RT \cdot [c + c' \cdot S_c \cdot \Delta G_{iwi}{}^{AB} \cdot f(d)], \tag{5.5}$$

**Table 5.2** Osmotic pressures of aquatic solutions of 10 and 60% (w/v) polyethylene oxide (PEO) of 200 and 20,000 molecular weight, in Mega Pascals (MPa)[a]

| PEO concentration = 60% (w/v) | PEO concentration = 10% (w/v) |
|---|---|
| A | B |
| PEO 200 | PEO 200 |
| Π van 't Hoff = 7.3 MPa | Π van 't Hoff = 1.22 MPa |
| Π Polar = 20.2 MPa | Π Polar = 0.09 MPa |
| Π Total = 27.5 MPa | Π Total = 1.31 MPa4 |
| C | D |
| PEO 20,000 | PEO 20,000 |
| Π van 't Hoff = 0.07 MPa | Π van 't Hoff = 0.01 MPa |
| Π Polar = 20.2 MPa | Π Polar = 0.09 MPa |
| Π Total = 20.3 MPa | Π Total = 0.10 MPa |

a The Π van 't Hoff and Π Polar have been derived, based on Eqs. (5.3) and (5.5).

where c is the dissolved polymer concentration in moles/L. However, in: $[c' \cdot S_c \cdot \Delta G_{iwi}^{AB} \cdot f(d)]$, $c'$ is the concentration of smaller PEO sections which interact practically independently. These PEO sections have a molecular weight of about five EO monomers, i.e., an Mw of about 200. Such sections repel each other with a free energy of $\Delta G_{iwi}^{AB}$. The strand width of PEO is about 0.46 nm (van Oss, 1994, 2006) and as two such strands when repelling one another remain crossed at a 90° angle, the $S_c$ value is $(0.46\ \text{nm})^2 = 0.2116\ \text{nm}^2$, for PEO strands of all lengths. (High Mw PEO strands are able to cross each other in multiple places, within one PEO strand, as well as between different PEO strands.) At closest approach, two such crossed PEO strands then repel each other with a free energy, $\Delta G_{iwi}^{AB} = +52.5\ \text{mJ/m}^2$ (van Oss, 1994, 2006) per crossing locus so that, at contact (i.e., at $d = d_0$), the total polar repulsive energy ($\Delta G_{iwi}^{AB} = -\chi_{iw}$) is equal to:

$$(52.5 \times 0.2116 \times 10^{-14})/kT = (52.5 \times 0.2116 \times 10^{-14})/4.05 \times 10^{-14}$$
$$= +2.73\ kT, \quad \text{at } 20\,^\circ\text{C}.$$

It is however only at the highest attainable PEO concentration in water, that a value for $\Delta G_{iwi}^{AB}$ of +2.73 kT can be reached. At lower concentrations, the distance (d) between mutually repelling strands, when crossing one another at 90°, becomes greater than $d_0$ and the exact distance, d, between two such crossing PEO strands determines the degree by which the value of $\Delta G_{iwi}^{AB}$ between two such loci has been diminished. This can be calculated using the extended DLVO approach, as indicated in Chapter 3; see the top equation given in Table 3.2.

[It should be noted that the absolute value of $\Delta G_{iwi}^{LW}$ for PEO (at contact) is about 14 times smaller than that of $\Delta G_{iwi}^{AB}$ and may therefore, as a first approximation, be taken as negligible, at all distances. Similarly, as PEO is electrically neutral, its $\Delta G_{iwi}^{EL}$ may also be neglected.]

In Table 5.2 the two parts of the osmotic pressure as well as the total Π values are shown for aqueous solutions of PEO 200 (Mw ≈ 200) and of PEO 20,000 (Mw ≈ 20,000), both at 60% (w/v) and at 10% (w/v) concentration. Table 5.2 serves to indicate the *van 't Hoff* as well as the *polar* parts of the osmotic pressure of water-soluble polymers of low and high molecular weights and at extremely high and relatively low concentrations, in aqueous solutions. In each of the four cases shown, the Π *van 't Hoff* and Π *Polar* values are given, as well as their sums total.

### 4.3.1 The Π van 't Hoff values in aqueous systems

The Π *van 't Hoff* values in all cases reflect only the number of whole molecules or macromolecules per L aqueous solutions. Compare the differences shown in Table 5.2 between the Π *van 't Hoff* values for PEO 200 (blocks A and B) and for PEO 20,000 (blocks C and D), both at c = 60% and c = 10%. Clearly, the PEO 200 and PEO 20,000 Π van 't Hoff values differ by a factor 100, as do their molecular weights. This is because the Mw of PEO 20,000 is 100 times greater than that of PEO 200, so that at the same concentrations there are 100 times fewer PEO 20,000 molecules per L solution than PEO 200 molecules.

### 4.3.2 The Π *Polar* values in aqueous systems

The approaches for calculating Π *Polar* are different for PEO 200 and PEO 20,000. With PEO 200, as well as with PEO 20,000 (blocks A and C) the Π Polar values are *both* 20.2 MPa at c = 60% and they are *both* 0.09 MPa at c = 10% (blocks B and D). This is because from a polar interaction standpoint the PEO molecules are each sub-divided into 100 units of 200 Mw each (i.e., into units of PEO 200-size), each one of which can individually repel other such sub-units, situated on either the same or on a different PEO 20,000 chain. Therefore, paradoxical though it might at first sight appear, at each given concentration (e.g., at 60%, in water) the Π *Polar* values are the same for PEO 200 as for PEO 20,000.

At c = 60% *both* PEO 200 and PEO 20,000 are close to the maximum of the aqueous solubility for PEO, mainly because beyond that concentration there is no further room for more (hydrated) PEO. (In other words, c = 60% is not really quite the same as a formal "solubility" in the surface-thermodynamic sense, as defined in Eq. (7.1).) Thus at a concentration of 60% (w/v) the molecules of PEO 200 as well as of PEO 20,000 are assumed to be forced to approach one another as close as $d = d_0$ (i.e., at "contact"), while undergoing a mutual repulsion and a crossed configuration, as well as a mutual repulsion energy of $\Delta G_{iwi}^{AB} = +52.5$ mJ/m$^2$ (van Oss, 1994, 2006).

As shown in Sub-section 4.2, above, at "contact" (i.e., at $d = d_0$), which is assumed to be the case at c = 60%, $\Delta G_{iwi}^{AB} = +2.73$ kT, so that there Π *Polar* = 20.2 kT. On the other hand, *at* c = 10%, the mutually repulsive, crossed PEO molecules, of 0.46 nm thickness, the (still crossed) PEO molecules (for PEO 200), or sub-sections of about 200 Mw (for PEO 20,000) are considerably farther apart, i.e., on average, at $d = 5 \times 0.46 = 2.3$ nm. Then, applying the extended DLVO approach by using the top equation of Table 3.2, the $\Delta G_{iwi}^{AB}$ value at crossed PEO molecules or strands is diminished by a factor $\exp[(0.157 - 2.3)/0.6] = 0.028$ times

**Table 5.3** Comparison between Π *Total* and Π *van 't Hoff*, based on the data given in Table 5.2

| C = 60% | C = 10% |
|---|---|
| A<br>PEO 200<br>Π Total = 3.77 times Π van 't Hoff | B<br>PEO 200<br>Π Total = 1.07 times Π van 't Hoff |
| C<br>PEO 20,000<br>Π Total = 290 times Π van 't Hoff | D<br>PEO 20,000<br>Π Total = 10 times Π van 't Hoff |

weaker, so that then $\Delta G_{iwi}^{AB} = +0.77$ kT, using the right hand term of Eq. (5.5), so that Π *Polar* = 0.09 MPa (see Table 5.2, blocks B and D).

### 4.3.3 The Π *Total* values in an aqueous system

The Π *Total* values given in Table 5.2 correspond closely with the Π *Total* values determined experimentally by Arnold et al. (1988) for PEO's of 150, 400, 6000 and 20,000 molecular weight, and for each of the various aqueous PEO solutions at different concentrations, from 10% through 60% (van Oss, Arnold et al., 1990; see also van Oss, 1994, Chapters V and XV; 2006, Chapters VI and XIX).

## 4.4 Conclusions regarding osmotic pressure

### 4.4.1 Osmotic pressures of linear water-soluble polymers

When hydrophilic polymers of low to medium molecular weights are dissolved in water at a *high concentration*, their total osmotic pressure can readily attain a value which is several hundred times greater than the classical Π *van 't Hoff*; see Table 5.3 (block C), i.e., the case of a 60% aqueous solution of PEO 20,000, with a ratio, R = [Π Total)/(Π van 't Hoff) = 290. Such large osmotic pressures, of almost 30 MPa, or approximately of the order of ≈300 atmospheres are not exceptional: they are easily attainable by a variety of water-soluble polymers, which are strongly polar or they would not be soluble in water. PEO is by no means an esoteric or exceptional example: it is widely available in a large array of molecular sizes and is used, for instance, as the major hydrophilic constituent of the vast majority of non-ionic surfactants. The mechanism of the strong mutual repulsion of PEO molecules when dissolved in water is solely a consequence of the high positive value of $\Delta G_{iwi}^{AB}$, where i stands for PEO.

### 4.4.2 Osmotic pressures of globular water-soluble polymers

Similar results as with PEO can be obtained with hydrophilic, water-soluble, electrically neutral, non-linear macromolecules, such as partly cross-linked dextran, or with moderately negatively charged highly water-soluble "globular" proteins such as mammalian serum albumin. In mammalian blood, serum albumin is quantitatively the most prevalent protein. Among other functions it plays an important role in maintaining the body's crucial colloid-osmotic pressure[1]). In the human blood

[1] In cattle a disease occurs which is usually only recognized at the slaughterhouse, just before impending slaughter, which is called "bovine hydrohemia," in which the concentration of blood serum albumin is severely diminished, caused

circulation the colloid-osmotic pressure is about 34 cm $H_2O$ (Altman and Dittmar, 1971), which is equal to 0.034 kg/cm$^2$ (or about one thirtieth of an atmosphere). At a serum albumin concentration of about 4% (Altman and Dittmar, 1971, p. 307) one arrives for human serum albumin at a Π *van 't Hoff* value of ≈ 0.014 kg/cm$^2$, so that about 59% of the total osmotic pressure provided by blood serum albumin is caused by Π *Polar*.

by a pathologically decreased albumin synthesis by the liver. In the slaughterhouse this disease is readily recognized by pervasive oedema (excessive hydration) of the animal's muscles and other tissues, so that the entire carcases are unusable and have to be discarded (van Oss et al., 1960; van Oss, 1962).

CHAPTER SIX

# The Interfacial Tension/Free Energy of Interaction between Water and Two Different Condensed-Phase Entities, i, Immersed in Water, w

## Contents



## 1. The $\Delta G_{1w2}$ Equation Pertaining to Two Different Entities, 1 and 2, Immersed in Water, w

The polar version of the $\Delta G_{1w2}$ equation is as follows:

*Interface Science and Technology*, Volume 16
ISSN 1573-4285, doi: 10.1016/S1573-4285(08)00206-8

$$\Delta G_{1W2} = \gamma_{12} - \gamma_{1W} - \gamma_{2W}$$

$$= 2[(\sqrt{\gamma_1^{LW}} - \sqrt{\gamma_2^{LW}})^2 - (\sqrt{\gamma_1^{LW}} - \sqrt{\gamma_W^{LW}})^2 - (\sqrt{\gamma_2^{LW}} - \sqrt{\gamma_W^{LW}})^2$$
$$+ \sqrt{\gamma_W^{+}}(\sqrt{\gamma_1^{-}} + \sqrt{\gamma_2^{-}} - \sqrt{\gamma_W^{-}}) + \sqrt{\gamma_W^{-}}(\sqrt{\gamma_1^{+}} + \sqrt{\gamma_2^{+}} - \sqrt{\gamma_W^{+}})$$
$$- \sqrt{(\gamma_1^{+}\gamma_2^{-})} - \sqrt{(\gamma_1^{-}\gamma_2^{+})}] \quad (6.1)$$

(see also Eq. (2.12)).

## 1.1 $\Delta G_{1w2}^{LW}$, or the apolar component of $\Delta G_{1w2}$

The apolar component of $\Delta G_{1w2}$, i.e., the Lifshitz van der Waals component, $\Delta G_{1W2}^{LW}$, of Eq. (6.1) can also be written as:

$$\Delta G_{1w2}^{LW} = -2(\sqrt{\gamma_1^{LW}} - \sqrt{\gamma_w^{LW}})(\sqrt{\gamma_2^{LW}} - \sqrt{\gamma_w^{LW}}). \quad (6.2)$$

Equation (6.2) shows that, contrary to the apolar part (term 1 of Eq. (5.1)), i.e., $\Delta G_{iwi}^{LW}$, which can only be negative (i.e., attractive), or zero, $\Delta G_{1w2}^{LW}$ can be negative (attractive), or zero, as well as positive (repulsive). It can easily be seen from Eq. (6.2) that $\Delta G_{1w2}^{LW}$ assumes a positive value when:

$$\sqrt{\gamma_1^{LW}} < \sqrt{\gamma_w^{LW}} < \sqrt{\gamma_2^{LW}} \quad (6.3A)$$

or when:

$$\sqrt{\gamma_1^{LW}} > \sqrt{\gamma_w^{LW}} > \sqrt{\gamma_2^{LW}}. \quad (6.3B)$$

In completely apolar systems, comprising two different polymers, dissolved in an apolar solvent, under conditions obeying the inequalities stipulated under either (6.3A) or (6.3B), above, when one replaces subscript, w, for water by subscript, L, for an organic liquid, such repulsions (where $\Delta G_{1L2}^{LW} > 0$) between two different dissolved polymers, 1 and 2 occur quite readily; see van Oss, Chaudhury and Good (1989); see also Chapter 2, Sub-section 1.4 and Eq. (2.7)). This type of phase separation is of the category designated by Bungenberg de Jong (1949) as "coacervation," or "simple coacervation;" see also van Oss (1989a) and Sub-section 2.5, below.

In water $\Delta G_{1w2}^{LW}$ does not readily become positive (i.e., repulsive) on account of the already quite low value of $\gamma_w^{LW}(=-21.8\ mJ/m^2)$, because few other condensed-phase materials have $\gamma_i^{LW}$ values that are significantly lower. However when one of the interacting surfaces is the water–air interface (whose $\gamma^{LW}$ value is zero), $\Delta G_{1w2}^{LW}$ is usually positive (i.e., repulsive) in all cases where $\gamma_i^{LW} > 21.8\ mJ/m^2$; see, e.g., Docoslis et al. (2000) and Chaudhury and Good (1985); see also Chapter 11, below. It should be noted, however, that when interactions occur in water, $\Delta G_{1w2}^{LW}$ is never the only significant component: $\Delta G_{1w2}^{AB}$ is also always active and it can, independently of the sign of $\Delta G_{1w2}^{LW}$, have a positive (repulsive) or a negative (attractive) value.

### 1.2 $\Delta G_{1w2}{}^{AB}$, or the polar component of $\Delta G_{1w2}$

$\Delta G_{1w2}{}^{AB}$, like $\Delta G_{1w1}{}^{AB}$ (see Chapter 5, Sub-section 1.2) can assume a positive (repulsive) or a negative (attractive) value and is quantitatively the most important component in influencing whether one has attraction or repulsion between entities 1 and 2, immersed in water. The only way to predict whether entities 1 and 2 will attract or repel each other when immersed in water is by doing the necessary (e.g., contact angle) measurements to obtain the data required in Eq. (6.1). Then, when $\Delta G_{1w2}$ is positive one may predict repulsion and when $\Delta G_{1w2}$ is negative one may predict that attraction will prevail and the precise degree of repulsion or attraction is expressed by the sign and the quantitative value of $\Delta G_{1w2}$, similar to the case of $\Delta G_{iwi}$; see Chapter 5, Sub-section 1.3.3.

### 1.3 Possible role of $\Delta G_{1w2}{}^{EL}$

It should be stressed that if one desires to predict whether attraction or repulsion will prevail between entities 1 and 2 when immersed in water, one should never neglect also to take $\Delta G_{1w2}{}^{EL}$ into account, in addition to $\Delta G_{1w2}{}^{LW}$ and $\Delta G_{1w2}{}^{AB}$; see Chapter 3.

In biological (mammalian) systems $\Delta G_{1w2}{}^{EL}$ as a rule tends to be very low and may usually be neglected, on account of the high ionic strength of most mammalian liquids. However, without actually having measured the $\zeta$-potentials of the surfaces in question, at full physiological ionic strengths, which allow the determination of $\Delta G_{iwi}{}^{EL}$ or $\Delta G_{1w2}{}^{EL}$, one cannot be sure that neglecting the $\Delta G^{EL}$ values is really warranted.

## 2. Examples of $\Delta G_{1w2}$ Interactions

### 2.1 Hydrophobic attraction between a hydrophobic and a hydrophilic entity, immersed in water

#### 2.1.1 Hydrophobic attraction between hydrophilic entities and hydrophobic surfaces

There are some important differences between interaction energies among identical entities, when immersed in water ($\Delta G_{iwi}$) and interaction energies among different entities, also in water ($\Delta G_{1w2}$). These are especially striking in the hydrophobic attraction of quite hydrophilic molecules by low energy (hydrophobic) surfaces, immersed in water. Taking for instance hydrated human immunoglobulin G (IgG), which has the following surface properties, at pH 7: $\gamma_{IgG}{}^{LW} = 34$, $\gamma_{IgG}{}^{+} = 1.5$ and $\gamma_{IgG}{}^{-} = 49.6$ mJ/m$^2$ (van Oss, 1989a, 1989b), which would make the $\Delta G_{iwi}$ value for that IgG to amount to +27.8 mJ/m$^2$ or, with an $S_c$-value of about 4 nm$^2$, that $\Delta G_{iwi}$ would represent about +27.5 kT of repulsion between two such IgG molecules, immersed in water, showing IgG to be quite hydrophilic and readily soluble in water. However when that same hydrophilic protein needs to be further purified, one would use, for instance, the reversed-phase liquid chromatography technique (see Sub-section 2.4.1, below), where IgG will readily bind hydrophobically to porous glass beads whose surfaces have been coated with octadecyl (C18) groups, because the $\Delta G_{1w2}$ value of the interaction between IgG and C18 in water amounts to about $-12.3$ mJ/m$^2$, or $-30.5$ kT. [In the case of *attraction* between non-spherical polymers the $S_c$ value for molecules such as IgG would be about 10 nm$^2$ because with attractions the $S_c$ value is greater than in the case of repulsions (see the $S_c$ value of 4 nm$^2$ mentioned above, in a case of repulsion) see also Chapter 7, Figure 7.1.] In summary, most hydrophilic polymers have a high positive $\Delta G_{iwi}$ value, so that they strongly repel one another, but are nonetheless strongly attracted, in water, to significantly hydrophobic surfaces. Only exceedingly hydrophilic surfaces, such as those of human serum immunoglobulin A (IgA) are not attracted to low-energy (hydrophobic) surfaces when immersed in water; see Sub-section 2.4.2, on hydrophobic interaction chromatography, below.

It is therefore erroneous to believe that only two hydrophobic entities attract each other when immersed in water: one hydrophobic and one hydrophilic entity usually also attract one another in water, albeit with a somewhat lower energy than is commonly seen with the attraction between two hydrophobic entities, immersed in water.

### 2.1.2 Hydrophobic attraction of hydrophilic cells (e.g., bacteria) *via* partition in hydrophilic/hydrophobic water–oil mixtures

There are two semi-quantitative measurement methods for estimating the relative hydrophobicity/hydrophilicity of cell (e.g., bacterial cell) surfaces, by partitioning the cells in an immiscible aqueous/non-aqueous combination of water and an apolar liquid ("oil"). In such systems, after the admixture of cells to such a water–oil mixture and shaking the mixture, after subsequent settling out of the two liquid phases, one measures the number of cells adhering to the water–oil interface, vs the number of cells remaining in suspension in the aqueous phase. The simplest version of this hydrophilic/hydrophobic approach (using, e.g., a water–hexadecane mixture) called MATH (microbial adhesion to hydrocarbons), was described by Rosenberg et al. (1980). A somewhat more sophisticated version of MATH, called MATS (microbial adhesion to solvents) uses *various different* water–oil mixtures, comprising apolar, Lewis basic and Lewis acidic organic solvents, in each case mixed with an approximately equal amount of water (Bellon-Fontaine et al., 1996). The latter authors demonstrated that with this approach one could not only obtain an estimate of the relative hydrophilicity/hydrophobicity of different bacteria, but also of the Lewis basic vs Lewis acidic nature of given bacterial strains, such as *Streptococcus thermophilus* and *Leuconostoc mesenteroides*; see also Chapter 2, Sub-section 3.1.

## 2.2 Hydrophilic repulsion between different hydrophilic entities, immersed in water

### 2.2.1 Aspecific repulsion among blood proteins, cells and other entities, immersed in water

The hydrophilic repulsion between *different* hydrophilic entities, in water, behave much like two identical hydrophilic molecules or particles: they tend to repel one another, as their free energies of interaction ($\Delta G_{1w2}$) usually are positive. In biological systems, e.g., in the human peripheral blood circulation we want to consider the solubility of blood plasma proteins and the stability of circulating blood cells. Here the polar repulsion energies between similar ($\Delta G_{iwi}^{AB}$) as well as between different hydrophilic entities ($\Delta G_{1w2}^{AB}$), comprising proteins as well as cells, are generally much greater than electrical double layer energies ($\Delta G_{iwi}^{EL}$ and $\Delta G_{1w2}^{EL}$). This is a consequence of the high ionic strength ($\mu = 0.15$) of the aqueous phase of mammalian blood (see Chapter 3, Sub-section 6.4). For the components with the highest $\zeta$-potentials there still are residual small $\Delta G_{iwi}^{EL}$ and $\Delta G_{1w2}^{EL}$ repulsions, but quantitatively speaking, even in these cases the positive values of $\Delta G_{iwi}^{AB}$ and $\Delta G_{1w2}^{AB}$ nonetheless predominate.

It is therefore due to the strong $\Delta G_{iwi}^{AB}$ and $\Delta G_{1w2}^{AB}$-driven repulsion energies that in the normal course of events mammalian blood serum proteins repel one another and thus remain soluble, and blood cells also repel each other, to keep their distance and remain in stable suspension in the blood circulation without clumping together.

### 2.2.2 Mechanism allowing blood elements to surmount the aspecific repulsion in order to engage in specific attractions between proteins as well as cells

The normally occurring aspecific repulsion mentioned in the preceding sub-section does not exclude the possibility for specialized white cells [e.g., polymorphonuclear leukocytes (PMN's) or neutrophils] to attach themselves specifically to other, "foreign" cells, such as invading microorganisms (bacteria or parasites) in order to engulf and destroy them, or for specialized proteins such as immunoglobulins (antibodies) to attach themselves to foreign antigens. PMN's represent the first line of defense against foreign invading cells (van Oss et al., 1975, p. 31). PMN's can overcome the $\Delta G_{1w2}^{AB}$-driven repulsion field by extruding long narrow pseudopodia with a small radius of curvature (van Oss et al., 1975, p. 112) which allows them to pierce that repulsion field and make specific contact with the invading cells and subsequently engulf and destroy them; see also Chapter 14.

The mechanism by which protuberances of a small radius of curvature succeed in surmounting the overall (mainly AB-driven) repulsion fields among hydrophilic particles, cells and proteins, immersed in an aqueous medium, is based on the smallness of their radius of curvature. It can be seen in Tables 3.1–3.3 that the radius, R of spherical entities is always proportional to *all three*: LW, AB and EL energies of interaction. For instance two smooth cells with a radius, R, will attract or repel another with an energy that is 100 times greater than two spherical bodies with a radius, r, when r is only 1/100th as large as R. And therein lies the advantage of cylindrical cellular protuberances or appendices with a small radius of curvature, because such narrow processes can easily pierce the repulsion fields of other hydrophilic bodies or cells in order to make specific contact with them, a feat which smooth, large particles, bodies or cells cannot accomplish. (See Chapter 3, Section 8 and Figure 3.1.) This also explains the need for the small-radius processes or spikes on the surfaces of most viruses (which serve to attach to living cells in order to facilitate entry in these cells, which allows them to procreate parasitically) and also on the surfaces of "activated" mammalian platelets, which allows them to become "sticky" and attach to sites where bleeding would occur if not plugged by platelets in a timely manner (see van Oss et al., 1975).

## 2.3 Advancing freezing fronts, causing a repulsion or an attraction, depending on the hydrophilicity or hydrophobicity of the immersed particles, cells or macromolecules

Advancing (hydrated) ice fronts in water at 0 °C are hydrophilic, with the following surface properties: $\gamma_{ice}^{LW} = 28.5$; $\gamma_{ice}^{+} = 10.5$ and $\gamma_{ice}^{-} = 42.1$ (all in mJ/m$^2$). Similarly, for water at 0 °C: $\gamma_w^{LW} = 22.8$; $\gamma_w^{+} = 19.0$ and $\gamma_w^{-} = 37.0$ (also in mJ/m$^2$) (van Oss, 2006, p. 96). Now, such a *hydrophilic* freezing (advancing) ice front *repels* hydrophilic solutes, polymers, particles or cells that are immersed in water at 0 °C. On the other hand, *hydrophobic* polymers, particles or cells that are dissolved or suspended in water at 0 °C are *attracted* to the advancing freezing ice front, *which then engulfs such hydrophobic entities as the ice front advances* (van Oss, Giese, Wentzek et al., 1992; van Oss, Giese and Norris, 1992).

### 2.3.1 Inorganic particles and advancing freezing fronts

The *engulfing* properties of advancing freezing fronts are easily demonstrated by filling a large, closed, plastic cylindrical tube with an aqueous suspension of particles, followed by placing the vertical cylinder with its bottom on a cold surface, refrigerated to about −40 °C. The cylinder should be enclosed in a 5 cm thick layer of a plastic foam insulation. When filled with an aqueous suspension of about 1 to 2% of partly hydrophilic clay particles (a montmorillonite, Swy-1) and when freezing started at the bottom of the cylinder which was standing on the cold surface, freezing occurred in the upward direction, vertically along the axis of the cylinder. When only the top 0.5 cm of the aqueous suspension had remained unfrozen the cylinder was removed from the cold source and examined. About half of the Swy-1 particles had then been pushed upward by the advancing ice front and remained visible on top of the ice, still in a stable aqueous suspension but significantly more concentrated than the initial suspension. The other half of the Swy-1 particles had been engulfed by the ice on its way up. This was apparently a separation between the less and the more hydrophilic clay particles, where the more hydrophilic ones made it all the way to the top, while remaining freely suspended in water. A second experiment was done under the same experimental conditions, but with a suspension of Swy-1 particles which had been made hydrophobic by treatment with a quaternary ammonium base, hexadecane trimethyl ammonium bromide (HDTMA). At the end of that experiment *all* the (now hydrophobic) HDTMA-treated clay particles had become engulfed in the lower fifth of the ice column (van Oss, Giese, Wentzek et al., 1992).

The surface properties of the untreated Swy-1 particles were: $\gamma_{Swy\text{-}1}^{LW} = 42.0$; $\gamma_{Swy\text{-}1}^{+} = 0.008$ and $\gamma_{Swy\text{-}1}^{-} = 30.2$ mJ/m$^2$. At 20 °C the $\Delta G_{iwi}$ value for Swy-1 was +2.3 mJ/m$^2$, which shows that Swy-1 is hydrophilic at room temperature. However, as a consequence of the fact that the surface properties of water at 0 °C are significantly different from its properties at 20 °C (see Sub-section 2.3, above), at 0 °C the value for $\Delta G_{iwi}$ of Swy-1 in water decreases to $-15.8$ mJ/m$^2$, which makes it modestly hydrophobic at the freezing point of water. The reason for this is that a major change occurs in the surface-thermodynamic properties of water upon cooling to the effect that its $\gamma^+/\gamma^-$ ratio is almost halved: decreasing from a value of 1.0 at 20 to 0.514 at 0 °C (due to a shift in $\gamma_w{}^+ = \gamma_w{}^- = 25.5$ mJ/m$^2$ at 20 °C, to $\gamma_w{}^+ = 19$ and $\gamma_w{}^- = 37$ mJ/m$^2$ at 0 °C).

When treated with HDTMA, the Swy-1 clay particles became coated with hexadecyl groups. (For hexadecane: $\gamma_{hex}{}^{LW} = 27.5$ mJ/m$^2$; $\gamma_{hex}{}^+ = \gamma_{hex}{}^- = 0$.) This made the mildly hydrophilic Swy-1 particles extremely hydrophobic, especially at 0 °C, with $\Delta G_{iwi} = -106.4$ mJ/m$^2$. Ice at 0 °C has a $\Delta G_{iwi}$ of $+1.24$ mJ/m$^2$ and thus is mildly hydrophilic. The average interaction energy between Swy-1 (untreated) and ice in water at 0 °C: $\Delta G_{1w2} = +0.2$ mJ/m$^2$, which is borderline repulsive, as the advancing ice front experiment also appeared to show. HDTMA-treated Swy-1, when immersed in freezing-point water, on the other hand, had a $\Delta G_{1w2} = -9.5$ mJ/m$^2$, which agrees with the complete engulfment which was observed.

These results point to a possibly useful remedy against the formation of "potholes" in road surfaces during the winter in Northern climates. Here small holes or imperfections in road surfaces, filled with loose grit, sand or clay particles (which usually are all hydrophilic) give rise to much bigger holes formed by the particle-pushing action of advancing ice fronts during the freezing phase. The fact that larger holes had thus been formed only becomes apparent after thaw has set in. A relatively simple way of preventing this pothole-digging effect caused by advancing freezing fronts, would be to spray loose grit, sand or clay particles residing in small holes or imperfections, with an aqueous solution of approximately 0.5% of a quaternary ammonium base. The aim of this is to transform the negatively charged hydrophilic particles into hydrophobic particles which then allows these hydrophobized grit, sand or clay particles to become *engulfed* by advancing ice fronts during freezing, thus preventing a digging or enlarging of initially small holes by the freezing process.

## 2.3.2 Effects of freezing on blood cells

Human blood cells, such as erythrocytes, leukocytes as well as platelets, all are quite hydrophilic and are therefore repelled by advancing ice fronts. This causes a problem, especially with the freezing of erythrocytes (i.e., red cells), which are by far the most frequently transfused cells and which often are frozen in order to be able to transfuse them at a later date than is normally permitted with non-frozen red cells, kept at +4 to +8 °C suspended in an anticoagulant liquid, which may only be transfused within 42 days post-donation. The difficulty is that by the end of the freezing process, the advancing ice fronts have pushed the cells as well as the dissolved salts into a very small compartment, in which most of the cells *and* the salts have become concentrated by then. This causes the frozen cells to be surrounded by large amounts of equally frozen salt ions. Upon thawing many of the cells thus find themselves suspended in a liquid medium with a high osmotic pressure, which tends to lead to haemolysis. One cannot pre-treat the cells with quaternary ammonium base-type surfactants the same way as the clay particles discussed in the preceding sub-section, because that would also lead to haemolysis and thus to becoming too toxic to be transfused in patients. However, even though one cannot easily modify the blood cells, one can change the liquid in which they are suspended.

In 1992 we demonstrated experimentally that whilst human erythrocytes (fixed with glutaraldehyde), suspended in water, were *repelled* by a slowly advancing freezing front, the same cells, suspended in a 50% (v/v) glycerol–water mixture, were *engulfed* by the advancing freezing front (van Oss, Giese and Norris, 1992). By the admixture of 50% glycerol one significantly decreases the value of term 5 of Eq. (5.1). This is because glycerol, at 20 °C has a $\gamma_{gly}^{+}$ of 3.9 and a $\gamma_{gly}^{-}$ of 57.4 mJ/m$^2$, which differ considerably from the corresponding values for water; at 0 °C the value for $\gamma_{gly}^{-}$ is even higher. Thus the hydrophobizing capability of a 50/50 glycerol/water mixture due to a decrease in the positive value of term 5 of Eq. (5.1) caused by the admixture of glycerol is much lower so that upon freezing, the red cells in such a mixture the cells are *attracted* by the advancing freezing front which causes them to be engulfed, which does not damage the cells at all. Although one subsequently needs to wash out most of the glycerol, remaining small amounts of glycerol are not toxic. However, after thawing the excess glycerol has to be removed gradually in order not to expose the red blood cells to unduly high osmotic pressures during the washing process with isotonic aqueous solutions.

Another condition favoring engulfment by advancing freezing fronts is fast freezing. It is only while freezing very slowly that one approaches the equilibrium conditions at which $\Delta G_{1w2}$ values can predict with some accuracy whether one is going to achieve engulfment of cells by the advancing ice front (i.e., when $\gamma G_{1w2} < 0$), or whether one will observe rejection of the cells by that front (i.e., when $\gamma G_{1w2} > 0$). It is, however, possible to produce engulfment of particles or cells when under equilibrium conditions one would still have a net repulsion between a (slowly) advancing ice front and given particles or cells by simply speeding up the velocity of the advancing freezing front. Thus, upon "fast" freezing, particles or cells that would normally be rejected under close-to-equilibrium conditions, can nevertheless become engulfed. The critical velocity ($V_c$) of the advancing freezing front that will result in complete engulfment, depends on the size and the nature (e.g., the surface tension) of the particles or cells that are to be engulfed. For instance, for glutaraldehyde-fixed human red cells, $V_c = 88$ μm/s; for human granulocytes (PMN's), $V_c = 31$ μm/s and for human lymphocytes, $V_c = 13.5$ μ m/s (Spelt et al., 1982). With fast freezing a lower concentration of cryoprotectant (e.g., glycerol) is needed for optimal cell survival than with slow (equilibrium) freezing (see, e.g., Meryman (1966); see also van Oss 1994, p. 358; 2006, p. 370).

It was thought at one time that the "surface tension" of cells could be determined for a given cell-type, by measuring the critical freezing velocity, $V_c$, for that type of cells, at which they became engulfed by an advancing freezing front (Spelt et al., 1982). However, this approach cannot be recommended. To begin with, in the 1982 paper, results were compared with cell surface tensions obtained with just water contact angles, from which their surface tensions were deduced by using a so-called "equation of state," now known to be erroneous (see Chapter 1, Section 5). Furthermore, even assuming that one could somehow determine a cell's surface tension ($\gamma_c$) by contact angle measurement with a single liquid, or by measuring a cell's critical freezing front velocity ($V_c$), there would be very little one can do with that knowledge. As shown in Chapter 5, one needs to know the $\gamma_i^{LW}$, $\gamma_i^+$ and $\gamma_i^-$ values (for all polar or apolar materials) and to derive their $\gamma_{iw}^{LW}$ and $\gamma_{iw}^{AB}$ properties from these, before one can determine their interaction energies with one another, when immersed in water, or their interaction energies with different materials (as treated in this chapter, above).

## 2.4 Chromatographic applications of hydrophobic interactions and their reversal

The principles of hydrophobic attraction and its reversal form the basis of a few important chromatographic separation and purification methods:

### 2.4.1 Reversed-Phase Liquid Chromatography

Reversed-Phase Liquid Chromatography (RPLC) is one of the High Pressure Liquid Chromatography or "HPLC" approaches, which are now for commercial reasons often alluded to as High *Performance* Liquid Chromatography. RPLC is especially useful in the separation and purification of proteins. To that effect columns are used which are packed with, e.g., porous glass beads that are made hydrophobic by treating them with octadecyl ($C_{18}$) groups. In water such $C_{18}$-coated beads bind hydrophobically to all but the most hydrophilic proteins. For the subsequent elution of the bound proteins one *decreases* the hydrophobizing capability of the aqueous medium by the admixture of water-miscible organic solvents with much lower AB values than the 51 $mJ/m^2$ of water, such as methanol, ethanol, ethylene glycol, or acetonitrile.

### 2.4.2 Hydrophobic interaction chromatography

Hydrophobic interaction chromatography is used for the separation or purification of exceedingly hydrophilic proteins, such as human immunoglobulin A (IgA), which are too hydrophilic to adsorb hydrophobically onto octadecylated ($C_{18}$) bead surfaces, so that one cannot use RPLC (see Sub-section 2.4.1, above). However, when *increasing* the hydrophobizing capability of water by the admixture of salt ("salting-out"), e.g., with 1 molar $(NH_4)_2SO_4$, IgA will adsorb onto $C_{18}$-treated surfaces (Doellgast and Plaut, 1976; see also van Oss, Moore et al., 1985). The IgA can subsequently be desorbed from the hydrophobic $C_{18}$ beads by means of a buffer gradient consisting of gradually decreasing $(NH_4)_2SO_4$ concentrations.

## 2.5 Polymer phase separation in water

### 2.5.1 Aqueous phase separation caused by a repulsion among different polymers dissolved in the same solvent (coacervation)

Molecules of an organic polymer, which are only subject to LW forces when immersed in an organic solvent, can only *attract* one another, with a $\Delta G_{iwi}^{LW}$ value that is negative or zero. Therefore, unless the $\gamma^{LW}$ values of such polymers are very close to that of the solvent, they can only be soluble by becoming strongly solvated in the solvent (see Chapter 7, Section 7). Two *different* organic macromolecules, 1 and 2, which are both dissolved in the same organic solvent under purely LW conditions, can *repel* each other by means of a net LW repulsion and separate out into two different phases, where one phase is rich in polymer 1 and the other phase is rich in polymer 2. The conditions favoring such a purely LW repulsion have been outlined in Chapter 1, Sub-section 4.1.

In *aqueous systems* net LW repulsions between two different dissolved polymers do not occur, because for such an event to take place the LW component ($\gamma^{LW}$) of one of the polymers should then be significantly less than the 21.8 mJ/m$^2$ of water and such polymers would not be soluble in water, due to the always-present hydrophobic attraction in water of $-102$ mJ/m$^2$, which any known organic polymer (that should be hydrophobic if it has a $\gamma^{LW} < 21.8$ mJ/m$^2$) would be incapable of surmounting with the help of its very limited hydration repulsion capability (term 5 of Eq. (5.1)). It is clear that all water-soluble polymers have to be strongly polar (i.e., very hydrophilic) and it is known that these have a $\gamma^{LW}$ value between approximately 26 and 45 mJ/m$^2$ (cf. van Oss, 1994, 2006). Thus, hydrophilic water-soluble polymers can only be water-soluble if such polymer molecules repel one another when immersed in water, mainly owing to the high positive value of term 5 of Eq. (5.1). This means that the $\Delta G_{iwi}$ value of *identical* water-soluble polymers has to be positive, which is indeed the case, see, e.g., Table 7.1 [see also van Oss and Giese (2004) and van Oss (2006)].

It should be noted that whilst two water-soluble polymers do not undergo an LW repulsion, one can nonetheless encounter LW-based repulsions in water between the *water–air interface* ($\gamma_{air} = 0$) and a dissolved solute with $\gamma_S > 21.8$ mJ/m$^2$; see Chapter 11, Section 3.

Now, when two *different* hydrophilic polymers, 1 and 2, are dissolved in water, $\Delta G_{1w1} > 0$ and $\Delta G_{2w2} > 0$, from which it follows that it is also true that $\Delta G_{1w2} > 0$, so that polymer 1 and polymer 2 also repel one another when dissolved in water. The reasoning for this is as follows: When polymer 1 is identical with polymer 2, $\Delta G_{1w2} = \Delta G_{1w1} = \Delta G_{2w2}$, so that there is the ordinary repulsion between identical polymers, i, which allows them to dissolve in water, as $\Delta G_{iwi} > 0$. However when polymers 1 and 2 are *different* and both are dissolved in water, the molecules of polymer 1 repel each other with a different free energy, $\Delta G_{1w1}$, than the molecules of polymer 2, which repel each other with a free energy, $G_{2w2}$. In such cases polymers 1 and 2 also tend to have different *densities*. Then, whilst *identical*, mutually repelling polymer molecules, dissolved in water, have no reason whatsoever to separate into different phases, two *different* polymers, 1 and 2, will repel each other with different energies, which will induce them to migrate into different liquid phases, helped by the fact that they also tend to differ somewhat in their specific densities. Thus in aqueous polymer solutions one can obtain as many phases as there are different dissolved polymer species; see, e.g., Albertsson (1986, pp. 13–15).

It should be noted on the other hand that in purely LW polymer solutions (e.g., in organic solvents) one can never get more than two phases, because if one adds a third polymer species, 3, to an LW solution of polymers 1 and 2, which are already separated into two phases, polymer 3 will either join the phase containing polymer 1, *or* the phase containing polymer 2, because of the laws governing net LW repulsions, outlined in Sub-section 3.3 of Chapter 3).

It should also be noted that in aqueous as well as in purely LW polymer phase separation systems, measurable phase separations involving different dissolved polymers only become observable when a certain minimum concentration is reached for each polymer. This is because at higher polymer concentrations the average distance between polymer molecules is decreased and at shorter interpolymer distances the repulsive energies between macromolecules are increased, see the extended DLVO approach treated in Chapter 3. Similarly, polymer phase separation is favored when higher molecular weight polymers are used, because with equal free energies of mutual repulsion per macromolecule pair, expressed in units kT, the total free energy of repulsion per polymer pair is also increased when such macromolecules' contactable surface areas, or their radii of curvature are increased. (For the definition of kT units, see Chapter 3, Table 3.3.)

The type of phase separation where two (or more) different dissolved polymer migrate to different compartments of the (usually aqueous) solution, has been named "coacervation" by Bungenberg de Jong (1949). However, aqueous polymer phase separation has more recently been designated as "aqueous partition," especially when the separation in two phases is utilized for the separation of added protein or cell mixtures, where the various proteins or cells migrate preferentially to one or the other phase compartment; see especially P.-Å. Albertsson (1986), who pioneered that separation approach.

The separation between various proteins or cells according to their preference for one aqueous phase or the other works following a similar mechanism as the phase separation in aqueous partition: a given protein molecule or cell will migrate preferentially to that phase which contains the highest concentration of the dissolved polymer species which repels that protein least. [It should be recalled that in aqueous two-phase partition both dissolved polymers as well as all added proteins or cells repel one another to different degrees; see the monographs by Albertsson (1986) and Zaslavsky (1995).]

### 2.5.2 Aqueous phase separation caused by the attraction between different dissolved molecules or macromolecules (complex coacervation)

Complex coacervation (Bungenberg de Jong, 1949) is the phenomenon whereby two different dissolved polymers (or one dissolved polymer and one lower molecular weight solute) form a complex with each other, upon which the complex thus formed migrates to (usually) the lower part of the aqueous medium, while remaining in solution. Complex coacervation is different from, but closely related to precipitation or flocculation (see also van Oss, 1991b; van Oss, 1994, 2006).

## 3. WATER TREATED AS THE CONTINUOUS LIQUID MEDIUM FOR $\Delta G_{1w1}$ AND $\Delta G_{1w2}$ INTERACTIONS

In all equations used in this book, pertaining to interactions occurring in water, water is treated as the continuous liquid medium and its surface thermodynamic properties are those that were quantitatively expressed in Chapter 2, Section 2. This does not mean that these surface thermodynamic properties of water are to be considered completely unaltered throughout an aqueous medium in which non-aqueous condensed phase apolar or polar materials have been immersed. Although beyond a fairly short distance of such materials from the water interface, the properties of the bulk water may be considered to be randomly homogeneous, within distances of about 1 nm or less from the water interfaces with such materials, the local orientation of the adjacent water molecules within the first 1 nm from the interface is often considerably less than random. Indeed, the non-random orientation of the water molecules in close proximity of other condensed-phase materials which are immersed in it (or even of air), plays a crucial role in the propagation at a distance of the mutual attractions or repulsion between such materials, when immersed in water. However, the influence which this altered orientation of vicinal water molecules exerts on the actions at a distance occurring in the aqueous medium, is rather accurately incorporated in the free energy equations pertaining to $\Delta G^{AB}$ and $\Delta G^{EL}$ vs distance interactions; see Chapter 3, Tables 3.2 and 3.3. On the other hand, the directly proportional or squared proportional action at a distance equations of $\Delta G^{LW}$ interactions are held to be valid regardless of slight differences in orientation of close vicinal water molecules; see van Oss, Giese and Docoslis (2001) and see also Chapter 3, Table 3.1.

CHAPTER SEVEN

# Aqueous Solubility and Insolubility

## Contents



## 1. The Solubility Equation

When it became necessary to measure interfacial tensions between organic liquids and water by other means than drop-shape or drop-weight approaches (see Chapter 4, Sub-section 3.2) we proposed the following solubility equation (cf. Eq. (4.5)):

$$-2\gamma_{iw} \cdot S_c = kT \cdot \ln s \tag{7.1}$$

*Interface Science and Technology*, Volume 16
ISSN 1573-4285, doi: 10.1016/S1573-4285(08)00207-X

(van Oss and Good, 1996; van Oss, Giese and Good, 2002), which is of course equivalent to (cf. Eq. (2.11)):

$$\Delta G_{iwi} \cdot S_c = kT \cdot \ln s, \tag{7.2}$$

where $S_c$ is the contactable surface area between two molecules, i, immersed in water, w, at the minimum equilibrium distance, $d = d_0$, and s is expressed in mol fractions; for the definition of kT, see Chapter 3, Table 3.3. As long as the solubility of i, however small, is known (see, e.g., the case of the aqueous solubility of octane; van Oss and Good, 1996), an accurate $\gamma_{iw}$ value can be obtained using Eq. (7.1). Equation (7.1) is especially useful for solutes with aqueous solubilities varying from very low, to about 0.1 mol fraction (=5.56 molar). This is because as a consequence of the exponential nature of Eqs. (7.1) and (7.2), at s = 0.5 mol fraction one already has the equivalence of 27.8 mol/L, which is difficult to attain with any solute with a higher molecular weight than that of water. For instance when s exceeds one mol fraction one reaches virtually infinite aqueous solubility (see Table 7.1 and van Oss and Giese, 2004).

The correlation of aqueous solubility with a solute's interfacial tension with water (Eq. (7.1)) is therefore more useful with solutes which have a known, finite solubility, than with extremely hydrophilic polymeric solutes with a positive $\Delta G_{iwi}$ value that corresponds to a quasi-infinite aqueous solubility (see Eq. (7.2)).

## 1.1 The contactable surface area ($S_C$)

The validity of Eq. (7.1) is readily tested with alkanes, of known aqueous solubility. The advantage of using alkanes is that non-polar organic liquids such as alkanes are the only ones with which one can legitimately measure their $\gamma_{iw}$ values by using drop-shape or drop-weight approaches [see Chapter 4, Section 2 and van Oss and Good, 1996]. Once $\gamma_{iw}$ is experimentally obtained for a given alkane by, e.g., drop-shape measurement and when the aqueous solubility of the alkane in question is also known, one can obtain its $S_c$ value using Eq. (7.1). After having thus obtained the $S_c$ value for a given alkane, one may equate that value with the $S_c$ of the alkyl group of, e.g., the corresponding alkyl alcohol and, by adding a further 0.10 $nm^2$ for the OH group to the $S_c$ of the alkyl group, one obtains the $S_c$ value for the complete alkyl alcohol (van Oss and Good, 1996).

It should be stressed that the *shape* of the solute molecules under study, as well as the *positive or negative status* of their $\Delta G_{iwi}$ play important roles in the determination of their $S_c$ values. For instance, if one has a long, ribbon-like, linear, *hydrophobic* polymer molecules, with a length of 15 nm and a thickness of 2 nm, then, as they *attract* each other when immersed in water (i.e., $\Delta G_{iwi} < 0$), their $S_c$ value would be equal to approximately $2 \times 15 = 30\ nm^2$, because two such molecules would tend to attract one another over most of their length and width; see Figure 7.1A. On the other hand, two linear *hydrophilic* polymer molecules of comparable dimensions would, when immersed in water, *repel* one another and would, because of that repulsion (i.e., $\Delta G_{iwi} > 0$) approach each other (without actually touching) at a 90° crossed position which would give rise to an $S_c$ value of only $2 \times 2 = 4\ nm^2$; see Figure 7.1B.

**Table 7.1** Aqueous solubilities and values of the free energies of interaction between two molecules, i, Immersed in water, w: $\Delta G_{iwi}$

| Solubility | | $\Delta G_{iwi}$ (kT) | |
|---|---|---|---|
| Mol/L | Mol fraction | | |
| | 244.7 | +5.5 | Human serum albumin with 2 hydration layers |
| | 148.4 | +5.0 | |
| | 68.7 | +4.23 | DEX T150[a] |
| | 15.6 | +2.75 | PEO 6000[b] |
| | 7.4 | +2 | |
| | 4 | +1.4 | |
| | 3 | +1.1 | |
| | 2 | +0.7 | |
| | 1 | 0 | |
| 27.78[g] | 0.5 | −0.69 | |
| 5.56 | 0.1 | −2.30 | |
| | | | Glucose[c]; sucrose[d] |
| $5.56 \times 10^{-1}$ | 0.01 | −4.61 | |
| $5.56 \times 10^{-2}$ | 0.001 | −6.91 | |
| $5.56 \times 10^{-3}$ | 0.0001 | −9.21 | Octanol[e] |
| $5.56 \times 10^{-4}$ | 0.00001 | −11.51 | |
| $5.56 \times 10^{-5}$ | 0.000001 | −13.82 | Octane[f] |
| $5.56 \times 10^{-6}$ | 0.0000001 | −16.12 | |

a s = 68.7 mol fr.; $\Delta G_{iwi}$ = +4.23 kT (van Oss and Giese, 2004).

b s = 15.6 mol fr.; $\Delta G_{iwi}$ = +2.75 kT (van Oss and Giese, 2004).

c s = 0.048 mol fr.; $\Delta G_{iwi}$ = −3.04 kT (Docoslis et al., 2000).

d s = 0.035 mol fr.; $\Delta G_{iwi}$ = −3.35 kT (Docoslis et al., 2000).

e s = 0.000,081 mol fr.; $\Delta G_{iwi}$ = −9.42 kT (van Oss and Good, 1996).

f s = 0.000,002,2 mol fr.; $\Delta G_{iwi}$ = −13.02 kT (van Oss and Good, 1996).

g This column is not extended further upward; the next entry above this one would already amount to 55.6 mol/L and except for water itself there are very few other compounds of which one would be able to confine 55.6 mol, at ambient pressure and temperature in 1 L. The aqueous solubilities given for hydrated human serum albumin and for dextran and poly(ethylene oxide) (PEO) simply indicate that these polymers are theoretically infinitely soluble in water, in practice mainly limited by excessive viscosity at high concentrations. From C.J. van Oss, Interfacial Forces in Aqueous Media, Taylor and Francis, Boca Raton, 2006, p. 299, with permission.

## 1.2 Contactable surface areas do not apply to *spherical* molecules or particles

When two *spherical* molecules approach one another as closely as possible, i.e., at $d = d_0 = 0.157$ nm (see Chapter 3, Sub-section 1.1), their "contactable surface area" is still zero. Thus, for spherical molecules, Eq. (7.2) has to be expressed differently as it needs to incorporate the radius, R, of the spherical molecules. It should first be understood that in Eq. (7.2), $\Delta G_{iwi}$ is expressed in units of mJ/m$^2$ (= ergs/cm$^2$), for the free energy of interaction between *two plane parallel plates*. When multiplying $\Delta G_{iwi}$ by the value of $S_c$ and dividing by the value of 1 kT ($=4.05 \times 10^{-21}$ Joules, or $4.05 \times 10^{-14}$ ergs, both at 293 K), one obtains $\Delta G_{iwi}$ in units kT (still for two plane parallel plates).

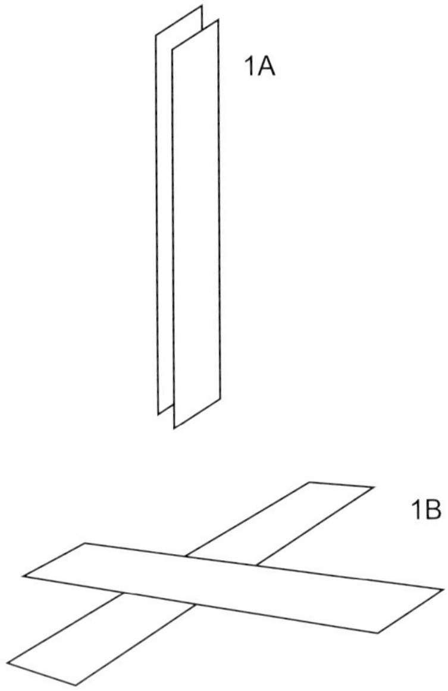


**Figure 7.1** Two ribbon-like molecules of $2 \times 15$ nm would have an $S_C$ value of about 30 nm$^2$ when they *attract* one another while immersed in water if they are *hydrophobic*; see the upper configuration (1A). On the other hand, two ribbon-like *hydrophilic* molecules of the same dimensions would be mutually *repulsive* while immersed in water and would therefore only be able to approach each other cross-wise, which causes them to have an $S_C$ value of only 4 nm$^2$; see the lower configuration (1B).

In order to transform a known value for $\Delta G_{iwi}$, expressed in mJ/m$^2$, in Eq. (7.2), into a value pertaining to two equal spheres, expressed, e.g., in Joules, one needs to change:

$\Delta G_{iwi}{}^{LW}$ (plane parallel surfaces) into $\Delta G_{iwi}{}^{LWoo}$ (two equal spheres of radius, R) by means of *multiplication* of $\Delta G_{iwi}{}^{LW}$ by: $\pi R/d_0$. (For an explanation of $d_0$, see Chapter 3, Sub-section 1.2.)
$\Delta G_{iwi}{}^{AB}$ (plane parallel surfaces) into $\Delta G_{iwi}{}^{ABoo}$ (two equal spheres, radius, R) by means of *multiplication* of $\Delta G_{iwi}{}^{AB}$ by: $\pi R\lambda$. (For an explanation of $\lambda$, see Table 3.2.)
$\Delta G_{iwi}{}^{EL}$ (plane parallel surfaces) into $\Delta G_{iwi}{}^{ELoo}$ (two equal spheres, radius, R) by means of multiplication $\Delta G_{iwi}{}^{EL}$ by:

$$0.5[\kappa \varepsilon R(\psi_0)^2]/[64n(\gamma_0)^2].$$

(For an explanation of these symbols, see Chapter 3, Table 3.3.)

To convert the S.I. answers into kT units one has to *divide* by the value of 1 kT, which is equal to $4.05 \times 10^{-21}$ J or $4.05 \times 10^{-14}$ ergs, at 20 °C (=293 Kelvin). All multiplication factors given above are for two equal spheres which are touching, i.e., approaching each other to a distance, $d = d_0 = 0.157$ nm (see Chapter 3). For interaction energies at a distance, at the two equal spheres configuration, use the appropriate formulae given in Tables 3.1–3.3. For the sphere/flat plate configuration one multiplies the "two equal spheres" values by a factor 2; see also Tables 3.1, 3.2 or 3.3.

# 2. Aqueous Solubility of Small Molecules

## 2.1 Aqueous solubility of small non-ionic organic molecules

Small, non-ionic organic compounds can be quite soluble in water at room temperature, without being infinitely soluble. For instance, the solubility of sucrose, $s = 1.95$ mol/L and for glucose, $s = 2.65$ mol/L, but such sugar molecules will nonetheless slightly attract one another while in aqueous solution. For instance, the free energy of attraction between sucrose molecules, dissolved in water, $\Delta G_{iwi} = -3.04$ kT and between glucose molecules, $\Delta G_{iwi} = -3.35$ kT. It is only when one approaches mutual attraction energies of about $-10$ kT per molecule pair that one begins to encounter low aqueous solubilities (see, e.g., octanol, at $\Delta G_{iwi} = -9.42$ kT, cf. Table 7.1), and extremely low aqueous solubilities at $\Delta G_{iwi}$ values that are even more negative than $-10$ kT (see, e.g., octane with $\Delta G_{iwi} = -13.02$ kT, cf. Table 7.1). Table 7.1 stops at aqueous solubility values for organic compounds that are smaller than $10^{-7}$ mol fractions because below that value such minuscule solubilities become too difficult to measure accurately, so that one finds few published solubility values that are even smaller.

## 2.2 Aqueous solubilities of inorganic salts

The aqueous solubility of inorganic salts is normally expressed post-dissociation into cations and anions upon their immersion in water (w). The cations then *mutually repel* one another electrostatically and so do the anions. However, when the cations (c) and the anions (a) *attract* each other more strongly, this counteracts solubility. The resulting residual free energy of attraction ($\Delta G_{cwa}$) is related to the net solubility in an analogous manner with small organic molecules, as described earlier in Eq. (7.2):

$$\Delta G_{cwa} \cdot S_c = kT \cdot \ln s. \tag{7.3}$$

Thus, once one knows the aqueous solubility, s, of a given salt in mol fractions, one can find the value for $\Delta G_{cwa}$, in kT units, by using Eq. (7.3).

Table 7.2 shows the solubilities at room temperature of five alkali chlorides, together with the values for $\Delta G_{cwa}$ (in kT) as well as the hydrated (Stokes) radii of the cations (Audubert, 1955). It can also be seen from the radii of the hydrated cations that increased hydration significantly decreases the slightly attractive $\Delta G_{cwa}$ values in the case of $Li^+$ and somewhat less in the case of $Na^+$ ions, but it makes very little difference among $K^+$, $Rb^+$ and $Cs^+$ ions, which have smaller and very similarly hydrated radii compared with $Li^+$ and $Na^+$. The heavier three alkali cations can clearly approach their counterion ($Cl^-$) more closely than $Li^+$ and $Na^+$ and thus attract it more strongly. In contrast, the stronger degree of hydration of $Na^+$ and

**Table 7.2** Solubilities and values of the free energies of interaction of alkali chlorides when immersed in water

| Alkali chloride | Aqueous solubility[a] (mol/L) | Aqueous solubility[a] (mol fraction) | $\Delta G_{1w2}$[b] (kT) | Hydrated stokes radius of the cation[c] (nm) |
|---|---|---|---|---|
| LiCl | 10.70 | 0.1926 | −1.65 | 0.232 |
| NaCl | 4.52 | 0.0814 | −2.51 | 0.176 |
| KCl | 3.42 | 0.0616 | −2.79 | 0.119 |
| RbCL | 3.95 | 0.0711 | −2.64 | 0.117 |
| CsCl | 3.87 | 0.0697 | −2.66 | 0.113 |

*a* From Stephen and Stephen (1963).

*b* Calculated according to Eq. (7.2). Even though cations repel cations and anions repel anions, it is assumed that the major factor in the aqueous solubility of alkali halides is the net residual *attraction* between cations (1) and anions (2), immersed in water (w). The smaller that attraction, the higher the aqueous solubility; see text.

*c* From Audubert (1955).

especially of $Li^+$ keeps them farther away from the $Cl^-$ ions, so that they are less attracted to their counterions and thus somewhat more soluble in water.

In principle one ought to be able to determine in this manner what the *total* free energy of interaction is between the cations and anions of salts which are soluble in water, once one knows their aqueous solubility. Thus far, however, there are too few data that would permit us to subdivide the aforementioned total free energy of interaction (i.e., the $\Delta G_{cwa}$ of Eq. (7.3), which for the present purpose may be more precisely described as $\Delta G_{cwa}^{TOT}$) into its $\Delta G_{cwa}^{LW}$, $\Delta G_{cwa}^{AB}$ and $\Delta G_{cwa}^{EL}$ components. We are further hampered by the fact that we also do not know the *ratio* of the *repulsive* free energies between cations as well as between anions, to the *attractive* free energies between cations and anions, while dissolved in water. Thus, whilst the aqueous solubility of non-ionic compounds leads directly to their interfacial tension ($\gamma_{iw}$) with water (see Eq. (7.1)), with electrolytes which dissociate into cations and anions when dissolved in water one can only determine the total net aggregate free energy of interaction ($\Delta G_{cwa}^{TOT}$), expressed in kT units. In addition, the $S_c$ value in Eq. (7.3) only has one single unique value when the electric potentials *and* the hydrated radii of both cations and anions have close to the same value (as, for instance, in the case of KCl; see Audubert, 1955). In actual fact the value of $\Delta G_{1w1}$, determined via contact angles measured on a flat face of a salt as well as of a sugar crystal appears to have no connection whatsoever with the $\Delta G_{cwa}$ value obtained from the aqueous solubility of the same salt. For instance the $\Delta G_{iwi}$ value obtained via contact angle measurements on a large NaCl crystal was found to be $+58.1$ mJ/m$^2$ (R.F. Giese and C.J. van Oss, 2006, unpublished results), whereas from the aqueous solubility of NaCl a $\Delta G_{cwa}$ value of $-2.51$ kT was found, see Table 7.2.

To summarize: the surface of the NaCl crystal is clearly quite hydrophilic: its surface tension properties are: $\gamma^{LW} = 42.6$ mJ/m$^2$; $\gamma^+ = 0$ and $\gamma^- = 68.4$ mJ/m$^2$, yielding a (positive) $\Delta G_{iwi}$ value of $+58.1$ mJ/m$^2$, whilst the (negative) $\Delta G_{cwa}$ value of NaCl, derived from its aqueous solubility equals $-2.51$ kT. This would suggest that when crystals of a salt such as NaCl are immersed in water, the crystals quickly dissolve first and only subsequently dissociate into cations and anions.

## 2.3 Aqueous solubility of surfactants manifests itself as their critical micelle concentration (cmc)

### 2.3.1 Surfactants

Surfactants are (usually) chain-like molecules of which the *tail* (i.e., the chain *sensu stricto*) is non-polar (i.e., hydrophobic) and one end terminates in a polar (i.e., hydrophilic) *head*. Surfactants are classified according to the physico-chemical nature of the hydrophilic head:

1. They are called *anionic* when the hydrophilic head is an anion, e.g., an $SO_4$ group as in sodium dodecyl sulfate, where the $SO_4$ end is the anion and the (in aqueous solution dissociated) sodium ion is its counterion. The dodecyl group then is its hydrophobic tail.
2. They are called *cationic* in the case of (usually) quaternary ammonium bases, e.g., trimethyl dodecyl ammonium bromide, where in aqueous solution the ammonium end is the cation and the bromium ion the (in aqueous solution detached) counterion.
3. They are called *non-ionic* in the case where one hydrophobic (e.g., alkyl) chain is attached to another chain-like, *non-ionic and hydrophilic* polymer, typically consisting of polyethylene oxide (PEO). With non-ionic surfactants the two chains often are of a comparable length, so that in these cases the image of a hydrophobic tail and a hydrophilic head is less descriptive of their structure than is the case with the ionic surfactants.

### 2.3.2 Surfactants do not precipitate, they form micelles instead

When surfactants, in aqueous solution, exceed the concentration at which the limit of their solubility is reached, they do not form insoluble clumps which precipitate, but the excess surfactant molecules organize themselves into structures consisting of closely-joined layers of hydrophilic heads (or hydrophilic chains, when non-ionic), exposed to the water interface, whilst the hydrophobic chains all stick together and point inward (away from the water), thus forming a *micelle*. This is a structure with *hydrophobic* chains interiorized whilst the hydrophilic moieties form the boundary which is in contact with the water. Thus the equivalent of *aqueous solubility* in the case of non-surfactant solutes, is called *critical micelle concentration*, or *cmc*, in the case of surfactants. The *cmc* is (by analogy with solubility), the concentration of the single dissolved surfactant molecules at the point where these molecules just start organizing together into micelles.

*Micelles* can be spherical, or cylindrical, or organized in multi-layered flat sheets. When a surfactant's aqueous solubility is exceeded, the resulting micelles may not be visible with the naked eye. Adamson (1990, p. 509) shows a variety of approaches by which the formation of micelles can nevertheless be demonstrated. These comprise, inter alia:

1. The gradual decrease in the surface tension of an aqueous surfactant solution, measured while increasing amounts of surfactant are being added, suddenly stops and even reverses, as soon as the first micelles are formed, i.e., when the surfactant's solubility is reached and the *cmc* begins to be exceeded.

2. The appearance of micelles, at the *cmc*, can also be followed by various approaches, such as light scattering, fluorescence quenching, or small angle neutron scattering.
3. The osmotic pressure of a surfactant, dissolved in water, increases in proportion with the concentration of added surfactant molecules. However, from the moment the *cmc* of the surfactant is reached, further increases in surfactant concentration cause only a very feeble increase in osmotic pressure. This is due to the fact that osmotic pressure is for an important part linked to the number of separate solute molecules and once such solute molecules begin to organize by the hundreds to form a single micelle, the number of separate immersed entities in the water increases at a much lower rate than before the *cmc* was reached.
4. With anionic or cationic surfactants, which are electrolytes, the beginning of micelle formation (at the *cmc*), is accompanied by a sharp increase in the solution's equivalent conductivity.
5. The fairly sudden appearance at the *cmc* of micelles which are quite large compared with the size of the initial surfactant molecules which formed them, causes a marked change in light scattering by the solution.

### 2.3.3 Stability of micelles

One could as easily discuss this in terms of the solubility of micelles, but given their size and heterogeneity, *stability* is here the preferred term. The stability of aqueous suspensions of micelles originating from anionic or cationic surfactants has long been well understood as being caused by the mutual repulsion between the cationic or anionic heads that find themselves at the water interface of the micelles, once these are formed. However, with *non-ionic micelles* the understanding of the stability of micelles in water has been more contentious. Nonetheless, the stability of non-ionic micelles is also a consequence of a simple repulsion, which in this case is a non-ionic repulsion, linked to a positive value of $\Delta G_{iwi}$ (see Chapter 5, Sub-section 3.1), where the positive sign of $\Delta G_{iwi}$ is largely lodged in term 5 of Eq. (5.1). The subject of non-ionic particle stabilization in aqueous suspensions is further treated in Chapter 8, Sub-section 2.2.

### 2.3.4 Correlation between the cmc and the surface-thermodynamic properties of the separate surfactant moieties

The correlation between the critical micelle concentration (cmc) of a given surfactant (S), when dissolved in water (W) may be expressed as:

$$\Delta G_{SWS} \cdot S_c = kT \cdot \ln \text{cmc}, \tag{7.3B}$$

where $\Delta G_{SWS}$ is equivalent to the $\Delta G_{iwi}$ given in Chapter 5, in the complete Eq. (5.1). It is generally quite feasible to express, for instance, the $\Delta G_{SWS}$ value for the apolar moiety in the guise of the alkyl group of a given surfactant by looking up the surface tension of its corresponding alkane in Jasper (1972). The surface tension of the polar moiety may have to be measured separately. However in *non-ionic* surfactants this is usually polyethylene oxide which, as a first approximation (for PEO 6000), has a $\Delta G_{iwi}$ value of $+52.5$ mJ/m$^2$ and an $S_c$ value of 0.21 nm$^2$, in the repulsive mode (van Oss, 1994, 2006). In Eq. (7.3B), $(\Delta G_{SWS} \cdot S_c)$ has to be expressed in kT units so that the cmc is then found in mol fractions. Also, the $\Delta G_{SWS}$ values for the apolar and the polar moieties of surfactants may only be added together after they have both been expressed in units kT.

With *anionic* and *cationic* surfactants the connection between the cmc value and the surface-thermodynamic properties of the constituting chains of such surfactants is somewhat more complicated. Here the $\Delta G^{LW}$ and $\Delta G^{AB}$ properties of the ionic heads must be determined separately from the $\Delta G^{EL}$ part. This can be done by coating an apolar, hydrophobic surface (such as a Teflon surface or a parafilm sheet) with the surfactant (dissolved in water), which makes the hydrophilic heads protrude away from the hydrophobic Teflon or parafilm surface. Then, after air-drying, the LW and AB properties of the hydrophilic heads can be determined via contact angle measurements.

The coating of hydrophobic latex particles with an appropriate ionic surfactant will result in particles that distally display the ionic, hydrophilic heads, so that they can be used in a micro-electrophoresis device for the determination of their $\zeta$-potential; see Chapter 3, Sub-section 5.2. From the measured $\zeta$-potential the $\psi_0$-potential follows (see Chapter 3, Sub-section 5.1 and Eq. (3.5)) after which Table 3.3 can be consulted to find the appropriate equation to obtain $\Delta G^{EL}$; see also van Oss and Costanzo (1992). Here also all $\Delta G^{EL}$ values have to be expressed in kT units before being added to the $\Delta G^{LW}$ and $\Delta G^{AB}$ values, in order to obtain the complete $\Delta G_{SWS}{}^{TOT}$, to be used in Eq. (7.3B), in compliance with $\Delta G_{SWS}{}^{TOT} = \Delta G_{SWS}{}^{LW} + \Delta G_{SWS}{}^{AB} + \Delta G_{SWS}{}^{EL}$.

# 3. Aqueous Solubility of Polymeric Molecules

## 3.1 Similarities between the aqueous solubility of polymer molecules and the stability of particle suspensions in water

To those who are familiar with analytical ultracentrifugation it is probably clear that whether one measures the sedimentation and diffusion coefficients of relatively globular polymer *molecules*, dissolved in water, or whether one obtains approximately the same coefficients with *small particles* which are suspended in water, the results thus obtained do not permit one to distinguish between dissolved molecules and suspended particles when entities of both categories are of approximately the same shape and size. This is because the rules pertaining to the aqueous solubility or insolubility of polymeric molecules are more comparable to those governing particle stability or flocculation in aqueous media than to the rules of aqueous solubility of small monomeric molecules. For polymeric molecules *and* for small particles, *solubility*, as well as *stability*, prevail when $\Delta G_{iwi} > 0$, i.e. when the polymer molecules, or the particles, repel one another when immersed in water. Small molecules on the other hand can be significantly soluble in water when their $\Delta G_{iwi} < 0$; see Table 7.1. However, in view of the many diverse properties of different classes of *water-soluble polymers*, their aqueous solubilities were judged to be more appropriately treated separately from the stability of aqueous particle suspensions, so that the latter have been relegated to a separate chapter, for which see Chapter 8, below.

## 3.2 Aqueous solubility of linear polymers

Water-soluble linear or ribbon-like polymer molecules will need to be hydrophilic (i.e., their $G_{iwi} > 0$) to remain soluble in water. Due to their size, with linear as well as with more globular polymer molecules any hydrophobic patch that is exposed to the water interface is likely to bind to another such patch on another similar polymer molecule, with a $\Delta G_{iwi}$ energy which is prone to amount to at least −10 kT, so that total insolubility then becomes inevitable. If any hydrophobic moiety still exists in a hydrophilic, water-soluble polymer such as polyethylene oxide (PEO) it must remain hidden from its interface with water as long as its structure permits. However, upon heating to temperatures >50 °C, its three-dimensional structure appears to change so that parts of its (hydrophobic) polyethylene backbone becomes exposed to the water interface, thus losing its aqueous solubility. The temperature at which this happens is called the $\theta$-temperature of the polymer in question. It should be obvious that with polymers which manifest a $\theta$-temperature above which they become insoluble, their aqueous solubility can only be meaningfully determined at temperatures below the $\theta$-value.

In Sub-section 1.1, above and in Figs. 7.1A and 7.1B, the influence of *attraction vs repulsion* on ribbon-like polymers, immersed in water are discussed and illustrated. In the *attractive mode*, hydrophobic polymers will have an $S_c$ *value* that can easily be an order of magnitude greater than is the case with a hydrophilic, water-soluble polymer of similar dimensions and shape, when in a *repulsive* mode.

## 3.3 Aqueous solubility of globular proteins

Most if not all mammalian plasma proteins are probably entirely soluble in the aqueous medium of the mammalian blood and lymph circulations. This is because "globular" proteins with a molecular weight ($M_w$) of about 50,000 or more must, to remain soluble at all, have a completely hydrophilic surface in contact with the water interface. The $\Delta G_{iwi}$ values of the surface of such proteins therefore have to be positive (see Table 7.1). To be designated as "globular" one may define that for oblong ellipsoids their asymmetry factor (i.e., their thickness divided by their length) should be greater than 1/10 so that their friction factor ratio $f/f_0$ should be less than 1.54. (The friction factor ratio of such an ellipsoid body ($f/f_0$) is the ratio of the friction factor of said body (f) divided by the friction factor of a spherical body ($f_0$) of the same (molecular) weight, cf. van Oss, 1994, p. 71; 2006, p. 76.) The friction factor (f) of a protein molecule is related to its diffusion constant (D) as:

$$f = RT/D, \tag{7.4}$$

where R is the gas constant ($R = 8.31 \times 10^{14}$ per degree K) and $f_0$ is:

$$f_0 = 6\pi\eta N(3M_w V\rho/4\pi N)^{1/3}, \tag{7.5}$$

where $\eta$ is the viscosity of the liquid, N is Avogadro's number ($N \approx 6 \times 10^{23}$), V is the partial specific volume (for globular proteins $V \approx 0.75$) and $\rho$ is the specific density of the solution.

A typical globular protein is human serum albumin (HSA). In the dissolved, hydrated state (with one layer of strongly bound water of hydration per HSA molecule) HSA has the following surface-thermodynamic properties: $\gamma^{LW} = 26.6$; $\gamma^{+} = 0.003$ and $\gamma^{-} = 87.5$ (all in mJ/m$^2$; see van Oss, 1994, 2006). HSA has a friction factor ratio, $f/f_0 = 1.288$ (Sober, 1968; p. C.10), which corresponds to an asymmetry ratio of $\approx 1/5.1$, if HSA may be considered to be a simple oblong ellipsoid. It actually has a V-shaped structure (He and Carter, 1992), but we shall nevertheless continue to regard HSA as having an oblong asymmetry factor of about 1/5, which still averages it as a globular protein with a contactable surface area ($S_c$) of approximately 3.0 nm$^2$ and with a $\Delta G_{iwi}$ value of about +63.5 kT which makes it infinitely soluble in water. However, in the normal dissolved state of HSA, it is only the outer (distal) peptides, situated at the water interface, that make it water-soluble. The tertiary structure of hydrated HSA is such that it also comprises a more hydrophobic "nucleus," which is normally completely surrounded by peptides consisting of hydrophilic amino acids when dissolved in water. However, when a native, hydrated HSA molecule encounters a hydrophobic surface, it tends to adsorb onto it by orienting its hydrophobic interior toward that hydrophobic surface. This happens also when HSA is being air-dried from an aqueous solution deposited on a flat solid surface, such as a glass microscope slide. It should be noted here that the water–air interface is the most hydrophobic surface known to Man (see Chapter 11, Sub-section 1.1). Thus, when air-dried, the once-hydrophilic HSA surface has metamorphosed into a hydrophobic one: its surface-thermodynamic properties are now: $\gamma^{LW} = 41.0$, $\gamma^{+} = 0.002$ and $\gamma^{-} = 20.0$ (all in mJ/m$^2$) (van Oss and Good, 1988; van Oss, 1989a, 1994, 2006), so that its $\Delta G_{iwi}$ value now is $-13.3$ kT.

This conversion of the hydrophilic globular protein, HSA, with a $\Delta G_{iwi}$ value of +63.4 kT per molecule pair, in the native, dissolved state, into a denatured, hydrophobic state after air-drying with a $\Delta G_{iwi}$ value of $-13.3$ kT seems to be an exceedingly drastic transformation into what appears to be a pronounced denaturation. Such air-dried, precipitated HSA particles float on water without at first dissolving. However, after some delay the white, dried HSA particles begin to turn gray and then dissolve fairly quickly thereafter and completely resume their original hydrophilic structure. Thus, the seemingly hydrophobic HSA molecules were only reversibly denatured following the air-drying process. All of the above-described events took place at pH 7. However, when one air-dries HSA from an aqueous solution at pH 4.8 (i.e., at the isoelectric pH of HSA) an even more drastic hydrophobization takes place: the dried HSA then has the following surface properties: $\gamma^{LW} = 44.0$; $\gamma^{+} = 0.03$ and $\gamma^{-} = 7.6$ (all in mJ/m$^2$), yielding a $\Delta G_{iwi}$ value of $-38.9$ kT. It should be noted that most globular proteins attain their lowest value of aqueous solubility when held at their isoelectric pH. It should also be noted that the $\gamma^{LW}$ value of *hydrated* HSA ($\gamma^{LW} = 26.2$ mJ/m$^2$) is much lower than that of dried HSA ($\gamma^{LW} = 41.0$ mJ/m$^2$). The latter value is more indicative of the $\gamma^{LW}$ of the pure HSA substance itself, whilst the 26.2 mJ/m$^2$ for *hydrated* HSA is, quite naturally, much closer to the $\gamma^{LW}$ value of water, i.e., 21.8 mJ/m$^2$.

Human immunoglobulin-G (IgG) behaves essentially like HSA upon air-drying (van Oss, 1994, 2006), although even in its native, hydrated state IgG is somewhat less soluble than HSA in water/ethanol mixtures, used for the fractional precipitation of blood serum proteins destined for transfusion and other medical applications (van Oss, 1989a).

The surface-thermodynamic properties of hydrated human IgG are: $\gamma^{LW}$ = 34.0, $\gamma^{+}$ = 1.5 and $\gamma^{-}$ = 49.6 (all in mJ/m$^2$), yielding $\Delta G_{iwi}$ = +27.8 mJ/m$^2$ which corresponds to +30.9 kT [estimating the $S_c$ of IgG at 4.5 nm$^2$; see van Oss, 1989a, 1989b, 2006] This is about half the value found for hydrated HSA (*vide supra*) but the main importance of these $\Delta G_{iwi}$ values is that they are both strongly positive so that HSA as well as IgG are completely (i.e., quasi-infinitely) soluble in water; see Table 7.1 and Section 1, above. Upon air-drying, IgG, like HSA, becomes significantly hydrophobic, for the same reason.

## 3.4 Aqueous solubility of non-globular, fibrous proteins

Non-globular, fibrous proteins may be defined as proteins that have an asymmetry factor (y) where y = $f/f_0$ (i.e., the ratio of thickness to length of an oblong ellipsoid, see Sub-section 3.3, above) is significantly smaller than 1/10. For example, the plasma protein, *fibrinogen*, with a friction factor ratio of 2.336 (Sober, 1968, p. C12), has an asymmetry factor y $\approx$ 1/29 (van Oss, 1994, p. 72; 2006, p. 77).

In contrast with HSA and IgG, *fibrinogen* remains as strongly hydrated after air-drying as it was in its initial, native state, when just coming out of aqueous solution (van Oss, 1994, p. 178; 2006, p. 219). This is best explained by the fact that fibrinogen (unlike HSA and IgG), being a long, skinny molecule, is 29 times longer than thick and, due to the constraints of its shape, is devoid of the interior hydrophobic peptides which in globular proteins tend to orient toward exterior hydrophobic surfaces, such as the water–air interface (see the preceding Sub-section 3.2, above). For bovine fibrinogen, as well as for bovine fibrin, surface-thermodynamic data exist (van Oss, 1990), showing that for bovine fibrinogen, $\gamma^{LW}$ = 42,6, $\gamma^{+}$ = 0.3 and $\gamma^{-}$ = 42.6 (all in mJ/m$^2$), yielding $\Delta G_{iwi}$ = +32.4 mJ/m$^2$ which, with $S_c \approx$ 6.0 nm$^2$, is equivalent to +48.1 kT and thus indicates that bovine fibrinogen is totally soluble in water. This bovine fibrinogen could be enzymatically converted to *fibrin*, by treatment with $CaCl_2$ and bovine thrombin (van Oss, 1990). The resulting fibrin had the following surface-thermodynamic properties: $\gamma^{LW}$ = 40.2, $\gamma^{+}$ = 0.3 and $\gamma^{-}$ = 12.0 (all in mJ/m$^2$), resulting in $\Delta G_{iwi}$ = −34.2 mJ/m$^2$, which, taking $S_c$ = 6.0 nm$^2$ (as in the original fibrinogen), yields $\Delta G_{iwi}$ = −50.8 kT. This shows that (bovine) fibrin is quite hydrophobic and virtually insoluble in water (cf. the data shown in Table 7.1).

## 3.5 Aqueous solubility of gel-forming polymers

An increased aqueous solubility at higher temperatures (T) occurs with many organic as well as inorganic solutes. In most of these cases this behavior is mainly a consequence of a change with T of the surface-thermodynamic properties of the solvent, *water*. The most important change as a function of T occurs in term 5 of Eq. (5.1) (see Sub-section 2.3 of Chapter 6, and especially, Section 2 of Chapter 9, below). With an increase in T, $\gamma_w^{+}$ increases and $\gamma_w^{-}$ decreases, so that whilst

$r = \gamma_w^+/\gamma_w^- = 1.0$ at 20 °C, at 38 °C, r = 1.75. Thus, at 38 °C, $\gamma_w^+ = 32.4$ mJ/m$^2$ and $\gamma_w^- = 18.5$ mJ/m$^2$. Therefore, as many water-soluble polymers (i) are monopolar or almost monopolar electron-donors with $\gamma_i^+ \approx 0$ and a fairly high $\gamma_i^-$ value (e.g., with $\gamma_i^- \gg 29$ mJ/m$^2$), the often already considerable value of term 5 of Eq. (5.1) further increases with an increase in T, because $\gamma_w^+$ increases with T. [It should be recalled that term 5 of Eq. (5.1) is: $+4\sqrt{(\gamma_i^-\gamma_w^+)}$]. Thus, this polar change in water as a function of T causes the positive value of $\Delta G_{iwi}$ to increase, or it actually can cause a negative $\Delta G_{iwi}$ to become positive. In the first case a water-soluble polymer can become even more soluble with an increase in T and in the second case biopolymers which are not soluble in water at room temperature can become soluble upon heating. Examples of the latter phenomenon are seen in the cases of gel-forming biopolymers such as gelatin and agarose.

The formation of a gel instead of, e.g., a powdery precipitate upon insolubilization of a linear polymer due to a decrease in T is caused by the propensity of specific sites on the strands of such ribbon-like polymers to attract each other via non-covalent ligand-receptor-like bonds. Due to the fact that such specific, mutually attractive sites are some distance apart, a porous, water-filled but otherwise insoluble network is formed following a decrease in T, which is habitually described as an aqueous *gel*. Most generally a powdery form of the gel-forming solute is dissolved in boiling water and the solution thus formed is allowed to cool down until insolubility is reached, upon which the solution rigidifies and the gel is formed, at a temperature of, say, about 40 °C. After this the gel is allowed to cool down further and is usually kept overnight at room temperature (≈20 °C). A significantly higher temperature than 40 °C is subsequently required for reliquification of the gel, i.e., for the gel-forming solute to reliquify through redissociation. This phenomenon is due to hysteresis of the gel-forming interpolymer chain bonds. Hysteresis is simply a strengthening as a function of time of the free energy ($\Delta G_{iwi}$) of binding at the discreet, specific binding sites between polymer chains, immersed in water. The mechanisms of hysteresis of non-covalent bond strengthening have been described earlier by van Oss, Docoslis and Giese (2001). These mechanisms can be two-fold: (1) Caused by a strengthening of an incipient bond [e.g., a hydrogen-bond, which is of the category of a Lewis acid–base (AB) interaction] through expulsion of interstitial water of hydration (see also Israelachvili and Wennerström, 1996) between the two binding sites, which alters the binding mode from the $\Delta G_{iwi}$ type to the $\Delta G_{iw}$ type, which is always stronger and thus enhances the bond and: (2) Through the decrease in distance (d) between existing but not yet engaged mutually attractive sites, so that the opportunity is favored for the attraction and subsequent binding between previously unapproachable neighboring sites [also described by van Oss, Docoslis and Giese (2001) and by Docoslis, Wu et al. (2001)]. Thus the aqueous solubility of linear, gel-forming polymers can in practice only be determined at, or *above* the gellification temperature. This is in contrast with the aqueous solubility behavior of polymers such as polyethylene oxide (PEO), whose solubility can only be determined up to or *below* the temperature of their $\theta$ point; see Sub-section 3.1, above. (The point of view that gel formation in water upon cooling may actually be more regarded as the association between two different but complementary moieties of the dissolved, usually linear polymers, is briefly addressed in Chapter 9, Sub-section 2.2.)

# 4. Influence of Temperature on Aqueous Solubility

As already mentioned in Sub-section 3.5, above, one important surface-thermodynamic property of water that changes markedly with a change in T is the $r = \gamma_w^+/\gamma_w^-$ ratio; see Chapter 9, Section 3. This ratio, r, which has been established at r = 1.0 at 20 °C, increases when T increases: at 38 °C, r = 1.75, whilst upon cooling to 0 °C, r = 0.51. Thus upon heating, $\gamma_w^+$ increases and $\gamma_w^-$ decreases, to the effect that especially term 5 of Eq. (5.1) increases in positive value (see also Sub-section 3.5, above), which causes polar (and especially monopolar electron-donating) materials to become more hydrated with an increase in T so that such monopolar or close-to-monopolar molecules, immersed in water, repel each other more strongly when the temperature is raised. Thus the total value of $\Delta G_{iwi}$ (Eq. (5.1)) which becomes less negative or becomes more positive with an increase in T, in turn enhances the aqueous solubility of polar molecules (i).

In conclusion, the surface-thermodynamic properties of water and especially its $\gamma_w^+/\gamma_w^-$ ratio strongly influence the aqueous solubility of polar solutes, as a function of temperature.

# 5. Aqueous Insolubilization (Precipitate Formation) Following the Encounter between Two Different Solutes that Can Interact with Each Other When Dissolved in Water

## 5.1 Classes of pairs of compounds that readily precipitate when encountering each other in aqueous solution

When two different solutes, dissolved in water, interact with one another and when that interaction gives rise to an insoluble compound, the resulting precipitate can display some unusual properties. Precipitates in aqueous media can arise through the encounter of *inter alia* (van Oss, 1994, p. 264; 2006, p. 289):

I. Acids and bases [e.g., $H_2SO_4$ and $Ba(OH)_2$].
II. Anionic and cationic surfactants.
III. Polycations and polyanions (e.g., gelatin and gum Arabic, at the appropriate pH where gelatin is positively and gum Arabic negatively charged).
IV. Electron acceptors and electron-donors (e.g., polyacrylic acid and polyvinyl pyrrolidone).
V. Antigens and antibodies.

The free energy of interaction in water in all five cases is expressed as $\Delta G_{1W2}$; see Chapter 6 and Eq. (6.1). Here precipitation occurs when $\Delta G_{1W2} \ll 0$ (in terms of $mJ/m^2$) or when $\Delta G_{1W2}$ is more negative than $-10$ kT.

## 5.2 Mechanism of insolubilization upon the encounter of two different compounds with opposing properties

All five categories of compounds with opposing properties that become insoluble in water when encountering each other in aqueous solution have one thing in common: Upon their encounter, their complementary polar moieties combine with one another because of opposing properties of one kind or another, which combination occurs at the expense of a local loss of polarity (usually through mutual neutralization of the reactive polar moieties). The resulting loss of polarity (and concomitantly, of hydrophilicity), causes the complex that is formed in this manner to change from being amply hydrophilic for each of the compounds to be soluble in water, to attain a degree of hydrophobicity which then causes it to become insoluble in water. In other words, the complex, once formed, changes from the original hydrophilicity and aqueous solubility of each of the separate partners, to hydrophobicity and concomitant insolubility of their complex, once formed.

The mechanisms of precipitate formation of categories I, II and III (Subsection 5.1, above) are quite similar, as they are all based on the mutual neutralization of oppositely charged moieties. Category IV uses a slightly different approach, i.e., the interaction between electron-donating and electron-accepting compounds (which can be electrostatically neutral). This mode of interaction was first proposed by Dobry (1948) but has thus far been less frequently utilized than the more obvious polymeric cation-anion neutralization method (category III). The latter approach was the principle used by Bixler and Michaels, 1969; see also Michaels, 1976; to insolubilize complementary polymers into gels consisting of a combination of a polyvinyl benzyl trimethyl ammonium salt with polystyrene sulfonate. These gels could be used as anisotropic ultrafiltration membranes tailored to feature an array of different pore sizes that were suitable for laboratory-scale separation, purification and concentration of proteins. These have since been replaced by membranes of a more recent vintage which can be found in, e.g., Fisher or Sigma catalogues.

Finally, precipitates formed by the interaction between antigens (Ag) and antibodies (Ab) (category V, see the preceding sub-section, above) can be caused by electrostatic (EL) interaction between, e.g., an acidic Ag such as double-stranded (ds) DNA and a more alkaline Ab specific to dsDNA. However a more frequently encountered pathway of antigen–antibody binding is that of hydrophobic (AB) bond formation, whilst the weaker LW bonds occur in all cases, once the Ag- and Ab-specific sites can approach each other closely. It should be recalled that all three interaction classes (LW, AB and EL) are of the weaker, non-covalent bonding variety: Ag–Ab bonds are virtually never covalent, but among the weaker bonds, the hydrophobic (AB) variety is the strongest. It should be recalled that in humans, the ionic strength of most fluids is approximately 0.15, which is elevated enough to be only compatible with rather weak $\zeta$-potentials.

It should be stressed that in mammalian and avian systems most antigenic determinants, called *epitopes*, are hydrophilic whilst the antibody-active sites (called *paratopes*) often are hydrophobic. The reason why most *epitopes* by far are hydrophilic is that to be recognizable as such they have to be situated at the water interface of antigenic biopolymers or cells and all *epitopic* sites that are placed at that water interface have to be hydrophilic, or else they will cause the antigenic biopolymer, particle or cell to clump hydrophobically when suspended in any aqueous liquid medium. This is why *paratopes* often are hydrophobic, in order to be able to bind (hydrophobically) to their corresponding *epitopes*. (It should be recalled that hydrophobic interactions can perfectly well occur between a hydrophilic and a hydrophobic site; see Chapter 6, Sub-section 1.1.) The hydrophobic attraction moieties on antibody-active sites on the Ab's (i.e., the *paratopes*) are hidden in a "cleft" of the Fab chains of the Ab, so that an Ab, in aqueous solution, will not interact and hydrophobically cause clumps with other Ab's that happen to be in their vicinity. With most specific Ag–Ab bond formations probably the crucial prerequisite is complementarity of the three-dimensional structures of *epitope* and *paratope*, to ensure the possibility of encountering one another over a maximum of their surface areas, to the minimum possible distance, where ideally $d \rightarrow d_0$.

In summary: *Epitopes* on the Ag's are prominently exposed and hydrophilic, whilst the *paratopes* on the Fab part of Ab's are hidden in a cleft, to prevent their hydrophobicity from causing undesirable clumping with other biopolymers or cells (see van Oss, 1998, 2003; see also Chapter 14, below).

## 5.3 Mechanism of the formation of specifically impermeable precipitate barriers

When precipitation between different molecules (1 and 2), dissolved in water is caused to take place in fairly large-pore gels (e.g., consisting of $\approx$1% agarose in water), in such a manner that dissolved molecules 1 and 2 each are placed in a different well in the gel, separated by about 1.0 to 1.5 cm, and are allowed to diffuse toward each other, they tend to form a precipitate line. This line is actually a sort of membrane, seen from above. That precipitate membrane then has the property of being impermeable to molecules 1 and 2, even as each one continues to diffuse toward the precipitate membrane, or barrier, which they formed. Even though they keep diffusing toward that barrier, once formed, it does not thicken further as long as some of the dissolved molecules 1 and 2 were initially present in their respective wells at approximately equivalent concentrations. Meanwhile the membrane or barrier remains permeable to other ions or molecules which are not related to or reactive with, the two initial molecules, 1 and 2 (van Oss, 1968 and van Oss and Heck, 1961).

The mechanism of specific impermeability to the forming molecules or ions is a very simple one: The barrier which forms very early on, when the first few molecules of types 1 and 2 encounter each other, being driven by diffusion, and form a precipitate. Subsequently, that early precipitate membrane (which, when formed inside a gel plate and seen from above, looks like a sharp precipitate line) *becomes self-repairing* and remains so as long as a low but sufficient concentration of its forming molecules 1 and 2, on either side of the barrier remains present in solution. When a hole occurs accidentally in the barrier, some of molecule 1 tries to diffuse through it, but these molecules immediately encounter some opposing molecules 2 from the other side, so that they precipitate together and in so doing, plug up the hole.

## 5.4 Examples of specifically impermeable precipitate barriers or membranes

### 5.4.1 Precipitated salt membranes: $BaSO_4$ and a few other barriers

With a category I precipitate formed by 0.025 M $H_2SO_4$ (pH 1.47) on one side and the same concentration of $Ba(OH)_2$ (pH 12.67) on the other side of a cellophane membrane (with ≈2 nm diameter pores) so that the acid and the base solutions differed by 11.2 pH points, a voltage difference between the two sides was established at ≈650 mV, which voltage difference could be maintained for weeks by the precipitate membrane, after observing a number of precautions, such as avoidance of $CO_2$ from the air, evaporation of water, etc. (Hirsch-Ayalon, 1956). (For a pH difference of 11.2, at room temperature, a voltage difference of about 651 mV could be expected.) Other ions such as $Cl^-$ and $K^+$ could freely diffuse through the $BaSO_4$ membrane.

Salt precipitate (ionic) membranes were first described by Pfeffer (1877): he made copper ferrocyanide membranes inside the porous wall of an earthenware jar by the diffusion of an aqueous copper sulfate solution toward potassium ferrocyanide (also in aqueous solution, one solution being inside and the other outside the jar, which was immersed in the outside solution). These copper ferrocyanide membranes inside the wall of such a Pfefferian jar had very small pores and could be used for measuring osmotic pressures caused by various solutes, but their specific impermeability was apparently not yet noticed. However Beutner (1913, 1920) first observed the specific impermeability of such membranes, with the copper ions on one side of the jar and the ferrocyanide on the other side, by measuring the electric potential difference between the solutions on either side of the precipitate membrane. Donnan and Green (1914) also observed the membrane potentials in a similar system, using parchment as the aspecific, porous carrier. (Donnan, 1924 subsequently went on to develop his theory of membrane equilibria; see also van Oss and Heck, 1961; van Oss, 1968.)

### 5.4.2 Formation of specifically impermeable barriers caused by double diffusion in gels

***A. Double diffusion in gels, or immunodifusion*** With a view to improving diagnostic and analytical techniques in the medical field of immunology and independent of physicochemical endeavors such as precipitation in porous bodies (Pfeffer, 1877) or in membranes (Hirsch-Ayalon, 1956, 1957) or in porous sheets of parchment (Donnan and Green, 1914), the formation of precipitate bands in gel plates using antigens (Ag) and antibodies (Ab) was pioneered by Ouchterlony (1949); see also Ouchterlony's (1968, book).

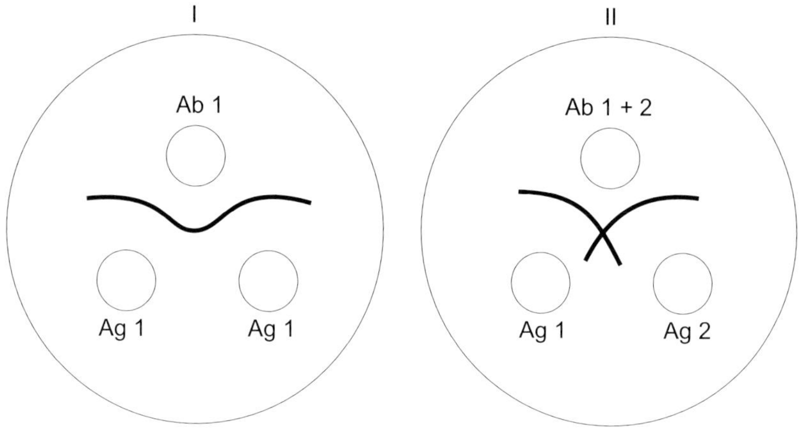


**Figure 7.2** Double diffusion precipitation lines formed by the interaction between an antigen (Ag) and an antibody (Ab) directed to the Ag. (I) In the top well is a solution of Ab1, directed against Ag1. Samples of the Ag1 solution are deposited in both lower wells and as in both these lower wells an *identical* Ag (i.e., Ag1) is deposited, the precipitate lines formed with Ab1 against both Ag1 samples, the two precipitate lines thus formed, will *fuse*, which is taken as a proof of *identity* between the contents of the two lower wells. (II) As in (I), but here there are two *different* Ag's: Ag1 and Ag2, whilst the Ab is a mixture of two Ab's, i.e., Ab1 against Ag1 and Ab2 against Ag2. As *these two systems have nothing in common*, the compounds making up the precipitate line formed by Ag2 having reacted with Ab2 can freely diffuse through the line formed by Ag1 having reacted with Ab1 and *vice-versa*, which allows the two precipitate lines to *cross* each other, which is taken as a proof of *non-identity* between Ag1 and Ag2, see text.

When molecules of an Ag and an Ab encounter each other by diffusing toward one another, each starting from a different well in an agar or agarose gel plate (e.g. a Petri dish), one sees from above the formation of a precipitate line, which is actually a precipitate membrane that is specifically impermeable to just the Ag and Ab molecules which formed it. The specific impermeability of the immunological precipitate lines formed upon the encounter between Ag and Ab molecules of various origins, furnished the explanation for the mechanism of the fusing or crossing of precipitate lines in various systems, connoting identity or non-identity, observed by Immunologists when using "Ouchterlony plates" (van Oss and Fontaine, 1961; van Oss and Heck, 1961). Ouchterlony (1949) had observed that when two Ag's, placed in two separate wells in a gel plate, upon encountering Ab's from an anti-Ag antiserum (comprising several different specificities) which were deposited in a third well, the two precipitate lines thus formed *fuse* when the Ag's were identical (Figure 7.2I) and *cross* when the two Ag's were unrelated (Figure 7.2II). The fact that the precipitate lines are actually precipitate barriers or membranes when seen from above and that they are specifically impermeable to just the Ag and Ab that formed them and hence fuse together when they are composed of *identical* compounds, was unknown to the many immunologists who started using this technique. When Ag1 and Ag2 are different, Ag1 and its Ab (i.e., Ab1) and Ag2 and Ab2 develop as separate specifically impermeable barriers, where still dissolved Ag1 and its Ab (Ab1) on

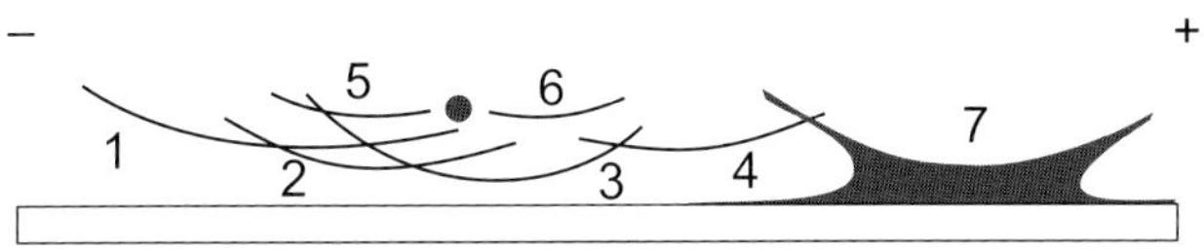


**Figure 7.3** Simplified presentation of immunoelectrophoresis of normal human serum proteins showing *six* of the maximally 32 serum proteins that can be visualized by this combined electrophoresis/double immunodiffusion technique, as selected from a normal human electropherogram (van Oss, 1979). This example serves to illustrate that the various different proteins which still are partly superimposed on one another after just the electrophoretic separation, can now be clearly distinguished from each other, thanks to the *crossing* of unrelated but neighboring precipitate lines produced by double immunodiffusion. Precipitate arcs of human serum proteins indicated by numbers are: 1 = IgG; 2 = IgA; 3 = transferrin; 4 = haptoglobin; 5 = IgM; 6 = Alpha2 macroglobulin; 7 = albumin.

the one hand and Ag2 and Ab2 on the other hand, combine as separate specifically impermeable barriers so that Ag1 and Ab1 can freely diffuse through the barrier formed by the reaction between Ag2 and Ab2 (and *vice-versa*). The result of this is that their different precipitates (and therefore their precipitate bands) can mutually interpenetrate, i.e., cross one another (Figure 7.2II) (van Oss and Fontaine, 1961; van Oss and Heck, 1961).

*B. Immunoelectrophoresis* Meanwhile the double diffusion precipitate system of crossing immunoprecipitate lines formed by unrelated Ag's reacting in gels with their Ab's was extended to the improved identification of many hitherto undifferentiated human blood serum proteins by the use of "immunoelectrophoresis" by Grabar and Williams (1953, 1955); see also Grabar and Burtin (1964). With this technique the proteins of whole human serum are separated into many separate fractions by a DC electric field, acting longitudinally across a rectangular gel plate, in the middle of which a (small) well was pierced, to serve as the starting reservoir for the sample of whole human serum. Then, in a second operation, a long narrow trough, parallel to the electrophoresed serum fractions, is filled with goat or rabbit antiserum to normal whole human serum, thus starting a double diffusion encounter between the electrophoretically separated human serum proteins (the Ag's) on the one side, and the various Ab's coming from the trough, which are directed against all the various normal human serum proteins; see Figure 7.3, where for clarity's sake only a small number of the curved precipitate lines thus obtainable are shown. What makes the various Ag–Ab precipitate lines identifiable as representing separate entities is the fact that they are curved because each of them originates from the double diffusion encounter between a single, small Ag source with an ellipsoid or circular shape, and its corresponding Ab which diffuses from a linear trough. It is the curved shape of each precipitate line which allows it to *cross* the nearest adjacent line formed by its closest but initially overlapping neighbor, thus permitting the identification of many more human serum proteins then was hitherto possible by simple electrophoresis.

This technique is still in use for the diagnosis of monoclonal gammopathies such as multiple myeloma (Kahler's disease). In such cases, instead of, e.g., *polyclonal* Immunoglobulin G (IgG) which is present in normal human serum at about 1% (w/v) and is seen as a long, almost linear, faintly curved, ski-shaped line (see Figure 7.3) because it consists of a whole family of IgG's comprising many different Ab specificities and a whole array of different electrophoretic mobilities, with a *monoclonal* (pathological) IgG one observes instead a short curved parabolic instead of a linear IgG fraction. In these cases such a curved, parabolic IgG fraction looks more like human serum albumin (HSA) (see Figure 7.3), but is still electrophoretically placed within the IgG mobility range. The immunoelectrophoretic detection of monoclonal IgG abnormalities can be further confirmed by doing a similar test, with pure (goat or rabbit) antihuman IgG antiserum, instead of antiwhole human serum, which makes the monoclonal IgG shape even more visible and which confirms that the pathological immunoglobulin is indeed of the IgG class. Similar tests can be done with antisera to the other immunoglobulin isotypes (i.e., IgA, IgD, IgE or IgM).

### 5.4.3 Precipitation in complex forming systems

Complex forming precipitate systems in water occur after the encounter of two complementary solutes which can combine in a non-stoichiometric manner at a wide array of ratios among the two solutes. Of the five categories of pairs of compounds which can precipitate when immersed in water, mentioned in Subsection 5.1, above, only Category I compounds, i.e., inorganic bases and acids (e.g., $Ba(OH)_2$ and $H_2SO_4$, forming precipitates of $BaSO_4$) form precipitates in a stoichiometric manner. The other four categories of precipitate forming compound pairs when immersed in water, all are *complex-forming compounds*: Category II: Anionic and cationic surfactants; Category III: Polycations and polyanions; Category IV: Electron acceptors and electron donors and: Category V: Antigens and antibodies. Pairs of compounds of these last four categories can all form precipitates consisting of widely varying non-stoichiometric proportions between the two reagents, when immersed in water and depending largely on the ratio of their respective concentrations at which they were initially present in aqueous solution. The term "complex-forming compounds" for this type of insolubilization was first suggested to me by Overbeek (ca. 1956).

***A. Influence of concentration differences*** Antigens (Ag) and their corresponding antibodies (Ab) are typical examples of complex forming systems that can form precipitates in water at a wide array of Ag/Ab ratios. Taking IgG isotype Ab's (IgG is divalent, i.e., it has two identical Ag-specific binding sites, or *paratopes*) and assuming a tetravalent Ag (i.e., an Ag molecule with four identical Ab-binding sites, or *epitopes*), Figure 7.4 illustrates the principal classes of complex formation as a function of the ratios of the concentrations of the participating Ag and Ab molecules initially present in aqueous solution. The tetravalent Ag is here represented as a sphere, whilst the IgG isotype Ab is shown with the shape of a real IgG molecule with its hinge wide open (see also Figure 14.1). Figure 7.4 depicts the two extreme conditions, of Ab-excess (*top*) and Ag-excess (*bottom*) with, in the *middle*, a complex which has formed at the "optimal" Ag/Ab ratio. In complex forming systems such as the Ag–Ab system "optimal" is not identical with "stoichiometric": One cannot usefully discuss the complex illustrated in the *middle* part of Figure 7.4 as

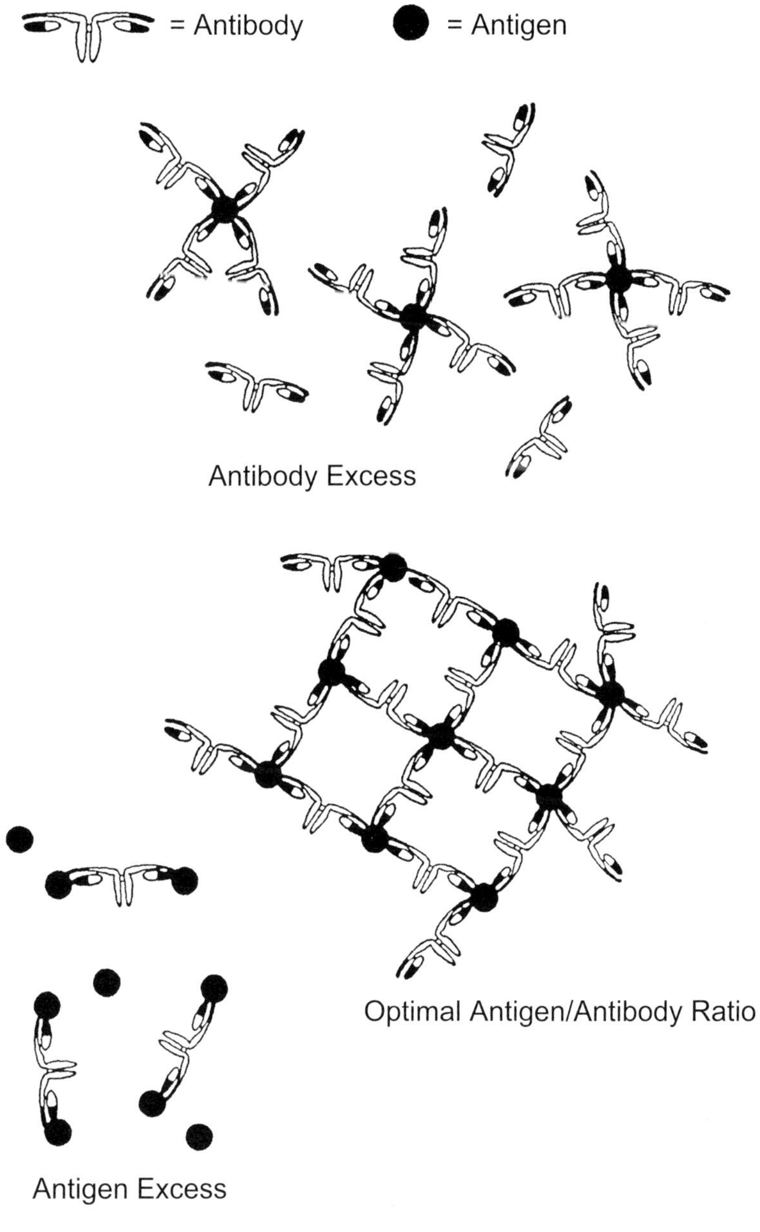


**Figure 7.4** Composition of three typical Ag–Ab complexes at different Ag/Ab ratios. The Ag is drawn as a small tetravalent (black) sphere and the Ab is a normal divalent antibody of the IgG isotype, with its hinge wide open, at 180°. *Top*: Complexes formed at Ab excess. *Middle*: Complex formed at optimal Ag/Ab ratio. *Bottom*: Complex formed at Ag excess (see text). From: C.J. van Oss [Precipitation and Agglutination, Figure 6.1, p. 82, in: N.R. Rose, F. Milgrom and C.J. van Oss (Eds.), Principles of Immunology, 2nd ed., Macmillan, New York, 1979, with permission of Scribner, imprint of Simon & Schuster Publishing Group, New York, dated March 9, 2007].

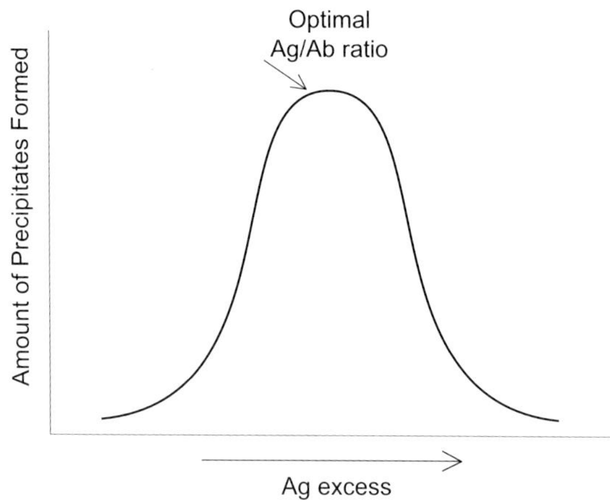


**Figure 7.5** Typical graph of the various amounts of Ag–Ab precipitate formed as a function of initial Ab excess, optimal Ag/Ab ratio and Ag excess (see text).

having occurred at the "stoichiometric" ratio; it is just the "optimal" ratio at which the greatest amount of precipitate is formed. For instance, at that optimal Ag/Ab ratio (here illustrated as $Ag_9Ab_{18}$) a semblance of stoichiometry might be apparent from the largest illustrated structure, but upon closer inspection there are still five available Ag valencies and six open Ab valencies. One can therefore only speak of optimal Ag/Ab ratios, linked to a maximum of precipitate formation and not to any stoichiometry, *sensu stricto*. It is also clear from the Ab-excess and Ag-excess examples shown in Figure 7.4 that in the smaller complexes formed in both these instances of higher proportions of exposed hydrophilic aminoterminals of the Fab moieties (at Ab-excess) and of the exposed hydrophilic epitopes (at Ag-excess), that the higher surface hydrophility of both types of smaller complexes favors increased aqueous solubility. (It should be stressed that the IgG *exterior* (outside the antibody-active site, or *paratope*, i.e., of the Fab moiety) is hydrophilic. The actual Ag-binding *paratope*, which is often at least partly hydrophobic, is internalized inside a cleft near the Fab extremity and does not influence the exterior hydrophilicity of the Fab part of IgG); see also Chapter 14, Sub-section 2.1.8, below.

The unusual solubility and precipitation properties of complex forming systems, of which Ag–Ab complex formation is one of the best known examples, are virtually exclusively seen in water. This is because it is mainly in water that the solubility of, e.g., biopolymers, is so strongly linked to their surface hydrophobicity. Thus the phenomenon of the formation of Ag–Ab complexes of differing solubilities according to the Ag/Ab concentration ratio of the Ag and Ab molecules present in solution, as illustrated in Figure 7.4, serves to explain the differences in behavior of the resulting Ag–Ab precipitate barriers, which are further discussed in this sub-section, B and C, below.

Figure 7.5 illustrates the degree by which the r = Ag/Ab ratio (indicated on the X-axis) influences the amount of Ag–Ab complexes formed (shown on the Y-axis). On the same curve the ordinate may also be taken to represent the molecular

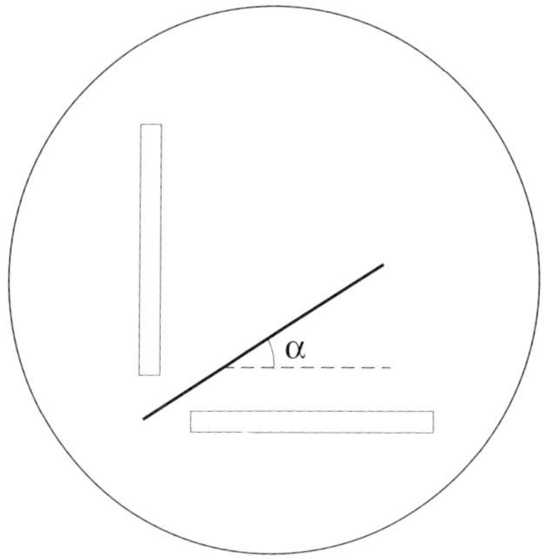


**Figure 7.6** L-plate used for the determination of the diffusion coefficient of given Ag molecules when immersed in water, when the diffusion coefficient of anti-Ag Ab molecules (also in water) is known; see Eq. (7.7) and text. (Neither the Ag nor the AB needs to be pure for this measurement.)

size of the insoluble complexes formed, as a function of r, as well as the relative aqueous solubility of these complexes, again as a function of r. Also, when the Ag–Ab precipitates are formed inside a carrier gel, the ordinate can also represent the percentage of the precipitate complexes that, once formed, can the least readily diffuse to other sites within the gel. However, in the presence of a large excess of one of the reagents (say, the Ag), even the largest precipitate complexes may alter their composition, resolubilize and move away from their original place of formation, to reform in a location where the ratio, r, of the two dissolved reagents is closer to unity.

*B. Place of first formation of the Ag–Ab precipitate lines formed by double diffusion in a gel* The place of first formation of a precipitate by a complex forming system such as an Ag–Ab precipitate formed by a double diffusion encounter in a carrier gel, is largely independent of the respective starting Ag and Ab concentrations (each originating from its own well, whose edges should not be much more than 1 or 2 cm apart). Under these conditions the place of first visible precipitate formation divides the distance, **x**, between the closest edges of the Ag and the Ab well as respectively **a** (from the Ag well) and **b** (from the Ab well) such that $\mathbf{a} + \mathbf{b} = \mathbf{x}$. In complex forming systems the $\mathbf{a}/\mathbf{b}$ ratio then is equal to (van Oss, 1968; van Oss and Heck, 1961; van Oss, 1984a):

$$\mathbf{a}/\mathbf{b} = \sqrt{(D_{Ag}/\mathbf{D}_{ab})}. \tag{7.6}$$

It should be noted, however, that with non-complex forming precipitate systems such as those involving inorganic salt precipitates (e.g., $BaSO_4$), the place of first precipitation obeys different and more complicated rules van Oss (1968) and van Oss and Hirsch-Ayalon (1959).

***C. Diffusion coefficients measured with L-plates*** One can make use of the constancy of the place of first formation of Ag–Ab precipitate lines in double diffusion in gels, which has been found to obey Eq. (7.6) for theoretical reasons and the correctness of which has been repeatedly experimentally confirmed (van Oss, 1968, 1984a), to measure the diffusion coefficient of Ag molecules via a double diffusion precipitation reaction with an Ab which specifically reacts with the Ag in question. Using an Ab that is an IgG isotype, its diffusion coefficient at 20 °C is known to be: $D_{ab} = 4 \times 10^{-7}$ $cm^2\,s^{-1}$. The most convenient configuration for measuring the value of $D_{Ag}$, by double diffusion precipitation with the corresponding Ab is an "L-plate", i.e., a gel plate in which two troughs are cut that are precisely perpendicular to each other, in the form of a capital "L", see Figure 7.6 (see Allison and Humphrey, 1960; see also Ouchterlony, 1949). Then, if the Ag solution is deposited in the vertical trough and the Ab solution in the horizontal one, the tangent of the angle ($\alpha$) the precipitate line makes with a line that is parallel to the horizontal trough, obeys the relation:

$$\tan\alpha = \sqrt{(D_{Ab}/D_{Ag})}. \tag{7.7}$$

As the value for $D_{ab}$ is known (see above) the value for $D_{Ag}$ is easily found (van Oss and Heck, 1961; van Oss, 1968). A unique advantage of the L-plate approach for measuring the diffusion coefficient ($D_{Ag}$) of a given antigenic molecule, is that the Ag in question does not have to be used in the purified state, as long as one does have an Ab (or even just an antiserum) which is specifically elicited against the Ag in question. Although the sharpest precipitate lines of first formation are formed when Ag and Ab are present at their optimal concentration ratio in the original starting solutions, the method still works when there is a fairly large departure from the optimal ratio (e.g., up to about a ten-fold discrepancy, either way).

***D. Titration of Ag or Ab concentrations by immuno-double diffusion*** The place of first formation of Ag–Ab precipitate lines (within the first 5 or 10 min to one hour from the start of the double diffusion) is remarkably constant and rather precisely predictable, once the diffusion coefficients of Ag and Ab are known (see Eq. (7.6)). However, the ultimate place where one finds the precipitate line after time intervals of several to many hours differs greatly from the predicted and observed place of first formation when the Ag/Ab concentration ratio deviates from the optimal ratio. This makes it possible to titrate, e.g., the original (unknown) Ab concentration *vis-à-vis* a known Ag concentration (or *vice-versa*). One of the simpler ways of effecting such a titration is illustrated in Figure 7.7. Here the optimal concentration of a bovine serum albumin/antibovine serum albumin (BSA/anti-BSA) is determined by titration via double diffusion in a gel plate with two rows of multiple wells. In the bottom five wells BSA is deposited in the following dilutions; from left to right: 0.5, 0.25, 0.125, 0.0625 and 0.0312%. In all the top five wells a goat anti-BSA antiserum, 50% diluted, is deposited. The precipitate line formation here illustrated shows that the thinnest line (which is the only line that has not moved up or down, or thickened, since its earliest visibility, at the beginning of the test) corresponds to 0.125% BSA, indicating that the ratio of the anti-BSA (the Ab) to BSA (the Ag) is as the original 50% Ab solution to 0125% BSA. If necessary a more precise Ag/Ab

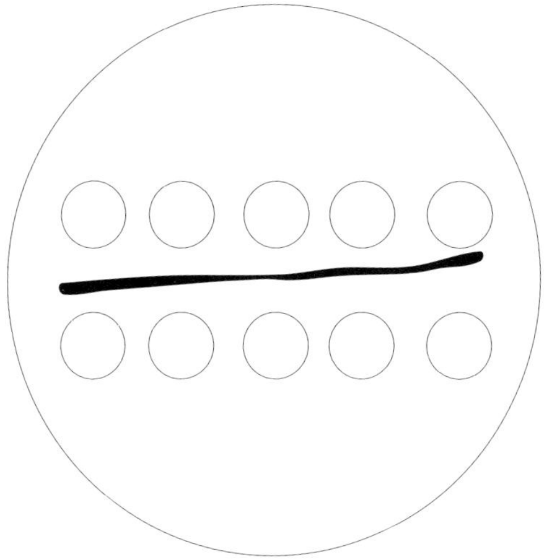

**Figure 7.7** Titration of a bovine serum albumin (BSA) concentration (the Ag), using an anti-BSA antiserum (the Ab). The Ab is 50% whole antiserum in all (bottom) wells and the top five wells contain BSA at the following dilutions (left to right): 0.5, 0.25, 0.125, 00625 and 0.0312%. The endpoint is at the optimal Ag/Ab ratio, which here can be seen to occur at 0.125% BSA, where one sees the sharpest and thinnest precipitation line, which has neither moved up nor down since its first formation. Here also neither Ag nor Ab needs to be present in a purified form (see text).

ratio can be determined by isolating the IgG from the goat anti-BSA antiserum and retitrating as above (see van Oss, 1984a). An analogous titration, using the same methodology as shown in Figure 7.7, was effected with another complex forming system, i.e., a sodium dodecyl sulfate vs hexadecyl trimethyl ammonium bromide (see van Oss, 1968), where it was observed that the double diffusion precipitation at equimolar concentrations of these anionic and cationic surfactants gave rise to the only precipitate line which did not move from its place of first formation, and persisted as the thinnest, sharpest line. Thus, in this case the optimal combination of an anionic with a cationic surfactant coincided with apparent stoichiometry, but the thickening of the other precipitate lines, together with the phenomenon of precipitate lines thickening and simultaneously moving away from the well containing the highest concentration of the surfactant of either sort, nonetheless is typical for complex forming systems.

## 5.5 Single diffusion precipitation

### 5.5.1 Single diffusion-driven precipitation in complex forming systems

Single diffusion precipitation is driven by the diffusion of reagent 1, present in aqueous solution, into a gel which has been homogeneously imbued with an aqueous solution of reagent 2, in such a manner that reagent 1 (in free solution) is significantly more concentrated than reagent 2 (present in the gel). Thus, when a finite, known amount of reagent 1 starts to diffuse into the gel which already contains the

**Figure 7.8** Single diffusion precipitation of an Ag, deposited as a fairly highly concentrated aqueous solution into a cylindrical well pierced in the middle of a round gel plate where the entire gel is imbued with a dilute Ab solution. From the moment of the start of the operation the Ag diffuses into the gel and immediately encounters dissolved Ab molecules, with which it begins to form a precipitate and, being much more concentrated than the Ab, the Ag quickly exhausts the Ab it finds in its path through precipitation, leaving the trapped precipitate particles behind in the gel. The Ag meanwhile continues to diffuse and while doing so, keeps encountering more Ab farther out in the gel with which it precipitates, until all the Ag is finally used up, at which point the circular precipitate disk thus formed stops growing. The amount of Ag deposited in the center well is then proportional to the surface area of the precipitate disk; see text.

much more dilute reagent 2, reagent 1 will immediately begin to form a precipitate with reagent 2. While diffusing further into the gel, reagent 1 exhausts all of the residual dissolved reagent 2 in its path, continuously binding more of reagent 2 which is imbedded in the gel and it will continue to advance and precipitate with reagent 2 until all the initial, known amount of reagent 1 is exhausted.

Typically a geometry is used in which a known amount of reagent 1 is deposited into a small cylindrical well placed in the middle of a (circular) Petri dish containing the gel imbued with dilute reagent 2 (Mancini et al., 1965); see Figure 7.8. In this configuration of single immunodiffusion, reagent 1 usually is the Ag solution whilst the gel is imbued with reagent 2, which is the Ab. The Ag molecules, while diffusing into the Ab-containing gel react and precipitate with the Ab molecules they encounter in the gel, but because the Ag is more concentrated than the Ab (although first forming a specifically impermeable barrier with that Ab) the Ag passes the barrier and reacts with new Ab further out, until all the Ag coming from the central well is used up. Once that happens the circular precipitate disk thus formed, does not grow any further. The surface area of the precipitate disk (or its diameter squared) then is proportional to the initial amount of Ag; see Figure 7.8. The entire process takes from one to several days, depending *inter alia* on the diffusion coefficient (i.e., the size) of the Ag molecules (Mancini et al., 1965).

### 5.5.2 Single diffusion-driven precipitation in non-complex forming systems—The *Liesegang* phenomenon

With non-complex forming, i.e., (generally) inorganic ionic systems, one can after some time observe the formation of multiple precipitate bands, or "rings," using a cylindrical single diffusion tube, usually filled with a gel (for stability) which is imbued with a dilute solution of one of the inorganic reagents (van Oss and Hirsch-Ayalon, 1959). The principal condition for the formation of multiple precipitate bands is to allow the diffusion of a concentrated aqueous inorganic salt solution (e.g., $PbNO_3$) into the (open) bottom part of a tube filled with a dilute gel (e.g., 2% agarose), which is imbued with a much more dilute salt (e.g., $Na_2CrO_7$), whereupon one ultimately observes an attractive looking array of multiple precipitate rings of $PbCr_2O_7$, i.e., *Liesegang* rings or band (see Liesegang, 1896, 1898). Stern (1967) compiled a bibliography on the subject, comprising 784 references, dating from 1855 to 1965. Many of the cited authors offer explanations for what came to be called the *Liesegang* phenomenon. These explanations are almost invariably wrong, as they "failed to take into account the (at least initial) specific impermeability of the precipitate bands and thus are likely to be at best flawed and more probably irrelevant and erroneous." What happens is that "once a precipitate barrier or band is formed at the meeting place of two reagents in a gel, that band remains specifically impermeable to the forming reagents until one of the reagents if exhausted (by precipitation and/or adsorption onto the precipitate), at least in the immediate vicinity of the precipitate band. From the moment the reagent that was initially present in the lowest concentration has become depleted in the immediate vicinity of the precipitate band, the most concentrated reagent is able to cross the barrier until it again encounters a sufficient concentration of the first reagent farther on, with which it will form a second precipitate band, etc." van Oss (1994). In further support of the above-mentioned mechanism of the *Liesegang* phenomenon of repeating precipitate bands, it should be mentioned that once the periodically recurring bands have been formed, the solution in the gel between two precipitate bands shows a complete, or almost complete absence of the reagent with which the gel was imbued and which was initially present at the lowest concentration; see, e.g.: Liesegang (1896, 1898); Bradford (1926); Hirsch-Ayalon (1956, 1957); van Oss and Hirsch-Ayalon (1959).

CHAPTER EIGHT

# Stability Versus Flocculation of Aqueous Particle Suspensions

## Contents



## 1. Stability of Particle Suspensions in Water

Part of the mechanisms governing the suspension stability of particles suspended in a liquid have been described by Derjaguin and Landau (1941) and by Verwey and Overbeek (1948), designated as the DLVO theory (see Chapter 3). This theory further elaborated upon principles laid out earlier by Hamaker (1936, 1937a, 1937b, 1937c).

The classical DLVO theory describes the outcome of electrostatic repulsion energies vs London–van der Waals attraction energies, as a function of interparticle distance, acting upon charged particles suspended in a liquid. As long as that liquid is non-polar the DLVO theory is correct. However in *polar* liquids and especially in *water* the classical DLVO theory only covers up to about 10% of the energies

*Interface Science and Technology*, Volume 16
ISSN 1573-4285, doi: 10.1016/S1573-4285(08)00208-1

governing the behavior of non-polar as well as of polar particles. To be descriptive of such systems as, e.g., the aqueous ones, it should be realized that Lewis acid–base (AB) interactions play the most crucial role and must be taken into account in addition to electrostatic repulsions and van der Waals attractions (see the extended DLVO theory, treated in Chapter 3). In all interactions occurring in water AB interactions play a principal and indispensable role, even in the case of suspensions of non-polar particles, where hydrophobic attraction ("hydrophobic effect") energies are *always* active and all by themselves cause an attractive energy between *all* particles (i) (and other objects) when immersed in water (w) such that $\Delta G_{iwi}^{AB\ (hydrophobic)}$ amounts to $-102$ mJ/m$^2$, at 20 °C (see Chapter 5, Section 2).

Nonetheless, this hydrophobic attraction energy, which is caused by the hydrogen bonding free energy of cohesion among the water molecules and is therefore always present in water, can be surmounted by an even stronger hydration repulsion (or "hydration pressure") energy exerted by hydrophilic entities when immersed in water (see Chapter 5, Section 3).

## 1.1 LW, AB and EL energies and the extended DLVO theory

### 1.1.1 LW energies in water

[For spherical particles or cells, the free energies of interaction, which in the following sub-sections are expressed in units of mJ/m$^2$ (i.e., free energies per unit surface area), often need to be converted into free energies per spherical particle pair of a given radius, R. When that needs to be done, consult Chapter 7, Sub-section 1.2].

Although the free energy of cohesion of water ($\Delta G_{ww}$) equals $-145.6$ mJ/m$^2$ (at 20 °C) and the Lifshitz–van der Waals part of this equals $-43.6$ mJ/m$^2$, or 30% of the whole, it must not be overlooked that the LW free energy of attraction between two particles, cells or molecules, immersed in water, $\Delta G_{iwi}^{LW}$ equals $-2(\sqrt{\gamma_i^{LW}} - \sqrt{\gamma_w^{LW}})^2$. Thus, for drops of for instance *octane*, immersed in water, $\Delta G_{iwi}^{LW}$ is virtually equal to zero, because for octane, $\gamma_i^{LW} = 21.6$ mJ/m$^2$ and for water, $\gamma_w^{LW} = 21.8$ mJ/m$^2$, so that $\Delta G_{iwi}^{LW} = -0.00092$ mJ/m$^2$ which is only 0.0006% of the free energy of cohesion of water and therefore entirely negligible. For octane in water that leaves only $\Delta G_{iwi}^{EL}$ (which is usually not enormous) as well as $\Delta G_{iwi}^{AB}$ which amounts to $-102$ mJ/m$^2$ (the absolute value of which is more than 100,000 times greater than $\Delta G_{iwi}^{LW}$).

For most other materials, immersed in water, $\Delta G_{iwi}^{LW}$ is usually not zero but about one or two orders of magnitude smaller than the always present hydrophobic term $\Delta G_{iwi}^{AB\ (hydrophobic)}$ of $-102$ mJ/m$^2$ and at most 10% of that value. (LW equations and measurement methods may be found in Chapter 2, Section 1; for LW equations pertaining to the DLVO approach, see Chapter 3, Section 3.)

### 1.1.2 AB energies in water

The AB free energies of interaction between (e.g., spherical) particles, immersed in water, can be found in Chapter 3, Section 4, which treats the role of AB interactions in the extended DLVO (XDLVO) approach.

The AB interaction energies, be they attractive or repulsive are, at $d = d_0$, quantitatively the most important ones for interactions between particles, immersed in water; they can range between $\Delta G_{iwi}^{AB} \approx -102$ mJ/m$^2$ for hydrophobic materials and $\Delta G_{iwi}^{AB} \approx +50$ mJ/m$^2$ or more for hydrophilic entities, see, e.g., Table 2.2, where the extreme negative as well as the extreme positive $\Delta G_{SWS}$ values are largely due to $\Delta G_{iwi}^{AB}$. Thus, given the relative exiguity of $\Delta G_{iwi}^{LW}$ in water as well as of $\Delta G_{iwi}^{EL}$ in biological and especially in human systems due to the high ambient ionic strength (see the next sub-section, below), the AB free energy of interaction typically accounts for at least 90% of the total $\Delta G_{iwi}^{TOT} = \Delta G_{iwi}^{(LW+AB+EL)}$, whether $\Delta G_{iwi}^{AB} > 0$ or $<0$.

### 1.1.3 EL energies in water

In biological and in particular in human systems, with an ionic strength, $\mu \approx 0.15$, which is mainly due to NaCl, the free energy of EL repulsion between, e.g., human erythrocytes (red cells) in normal human blood at close to ionic strength 0.15 and at ≈37 °C, is relatively small, compared to the AB repulsion between these cells under *in vivo* conditions. This exiguity of $\Delta G_{iwi}^{EL}$ *in vivo* is mainly a consequence of the high salt content of human blood. In the next sub-section a comparison is made between the LW, AB and EL interactions between human erythrocytes *in vivo*, noting also that although their $\zeta$-potential is fairly low, they still have the highest $\zeta$-potential of all human blood cells.

### 1.1.4 Comparison between LW, AB and EL contributions to the stability of erythrocytes in human blood under close to *in vivo* conditions

In the 1980's experimental proof emerged, using Israelachvili's (1985, 1991) force balance, which clearly showed that in water "non-DLVO forces" existed in addition to van der Waals attractions and electrostatic repulsions [see, e.g., Claesson (1986) and Grasso et al. (2002); see also Chapter 1, Sub-section 2.1]. In 1990 van Oss, Giese and Costanzo concluded from their measurements of the surface and electrokinetic properties of aqueous suspensions of the montmorillonite, *hectorite*, that the stability vs. flocculation properties of these clay particles were only quantitatively explicable when one included the influence of Lewis acid–base (AB) forces, in addition to the LW and EL forces (van Oss, Giese and Costanzo, 1990). An important biological example can also be utilized to clarify the various influences of LW, AB and EL interactions, involved in the suspension stability of human erythrocytes under close to the *in vivo* conditions which normally occur in circulating whole human blood.

In Table 8.1 the extended DLVO results are shown for human red cells, based on existing data on the surface and electrokinetic properties of these cells under *in vivo* conditions. To begin with, if one were to look at the free energies acting on the red blood cells, suspended in water, at the pH and ionic strength, as well as the approximate temperature of whole human blood, *without taking the AB energies into account*, one would conclude that human erythrocytes would be unstable at all intercellular distances, which obviously is not the case. If one only considered the sum of the LW + EL energies, as in the classical DLVO approach, $\Delta G_{iwi}^{TOT}$ would be, reading from d = $d_0$ through d = 20, 50, 60, 70 to 80 Å: −661, −47.5, −59.5, −0.6, −0.77 and −0.7 kT, respectively, which all denote an attraction, i.e., flocculation, which would be of an overwhelming impact at all distances up to d = 50 Å. It

**Table 8.1** Free energy, i.e., $\Delta G_{iwi}$, in kT units, between human erythrocytes, immersed in water, at ≈35 °C; ionic strength, $\mu = 0.145$ and pH 7.2, as a function of distance d between cells for LW, AB and EL interactions[f]

| $\Delta G_{iwi}$ (in kT) | Distance (d) | | | | | |
|---|---|---|---|---|---|---|
| | $d = d_0$ (=1.57 Å) | 20 Å | 50 Å | 60 Å | 70 Å | 80 Å |
| $\Delta G_{iwi}^{LW}$ | −1979[a] | −155.4 | −62.1 | −1.36[d] | −0.99[d] | −0.76[d] |
| $\Delta G_{iwi}^{AB}$ | +206,800[b] | +9583.9 | +64.6 | +12.2 | +2.30 | +0.44 |
| $\Delta G_{iwi}^{EL}$ | +1368[c] | +107.9 | +2.6 | +0.76 | +0.22 | +0.06 |
| $\Delta G_{iwi}^{TOT e}$ | +206,189 | +9536.4 | +5.1 | +11.6 | +1.53 | −0.26 |

*a* From data in van Oss (1994, 2006), adjusted to 35 °C; see equation for two equal spheres given in Table 3.1.

*b* As in [b] (see above), with equation for two equal spheres in Table 3.2.

*c* From electrophoretic mobilities of human erythrocytes, at 35 °C, $\mu = 0.145$ and pH 7.2, one obtains $\zeta = -10.9$ mV and $\Psi_0 = -15.8$ mV (see Chapter 3, Section 5). The $\Delta G_{iwi}^{EL}$ values at $d > d_0$ were calculated using Table 3.3, for the case of two equal spheres.

*d* $\Delta G_{iwi}^{LW}$ is much reduced at d > 50 Å because at d > 50 Å van der Waals–London retardation sets in; see Chapter 3, Sub-section 3.2.

*e* It will be seen that from $d = d_0$ to d = 70 Å the erythrocytes repel one another, thus guaranteeing their stable suspension in peripheral blood in close to normal *in vivo* conditions, which is largely a consequence of the $\Delta G_{iwi}^{AB}$ values. The value for the characteristic length for water at 20 °C, $\lambda = 6$ Å, is the value for water at 20 °C, but has been retained because the value for $\lambda$ at 35 °C (which is probably slightly less than 6 Å), is not known; for the role of $\lambda$, see Chapter 3, Table 3.2. It can also be seen that if one depended on the classical DLVO theory, using only $\Delta G_{iwi}^{LW}$ and $\Delta G_{iwi}^{EL}$, red blood cells would clump together at *all distances*, from $d = d_0$ to d = 80 Å; see Sub-section 1.1.4, above.

*f* In view of the somewhat unwieldy doughnut-like shape of human erythrocytes, their averaged radius has been estimated at $5 \times 10^{-4}$ cm.

is only thanks to the hydrophilic (positive, i.e., repulsive) $\Delta G_{iwi}^{AB}$ energies (which are *at contact* already 104.5 and 151.2 times stronger than $\Delta G_{iwi}^{LW}$ and $\Delta G_{iwi}^{EL}$, respectively) that the actual repulsive free energy of the very hydrophilic red blood cells prevails up to past d = 70 Å, beyond which (e.g., at d = 80 Å) only slightly attractive but ineffective total energies occur with a (negative) value of −0.26 kT and less at even greater distances. Furthermore, at all distances, $\Delta G^{EL}$ is smaller than $\Delta G^{LW}$, which is a consequence of the for EL energies unfavorable limitations of the biological *in vivo* conditions prevailing in human blood, especially with a view to its high ionic strength. Moreover, the difference in behavior between LW forces, in water, and AB forces, also in water, further contributes to make the value of $G_{iwi}^{LW}$, even at contact, appear rather small. This is further accentuated by the fact that at distances greater than 50 Å, LW forces decay much more steeply as at that point the London–van der Waals retardation sets in (see Chapter 3, Sub-section 3.2). Thus it is practically only owing to the red cells' pronounced surface hydrophilicity (caused by AB forces) and their concomitant strong mutual repulsion when immersed in water, that these cells remain exceedingly resistant to flocculation *in vivo*. Leukocytes also form stable suspensions in blood and in other biological fluids, largely following the same mechanisms as erythrocytes, given their fairly comparable surface properties (van Oss, 1994, 2006).

# 2. Stability of Charged and Uncharged Particles, Suspended in Water

Much has been written about the aqueous stability of suspensions of electrically charged particles; see, e.g., Overbeek (1952, pp. 302, ff.), treating inorganic "hydrophobic" particles immersed in water and the influence of London–van der Waals and electrical double layer interactions thereupon. In 1952 Lewis (AB) energies were not yet implicated in these interactions, but one may expect in the light of more recent studies (see, e.g., Section 3, below) that in reality these inorganic particles, at their pH of maximal stability were in effect, at that point, actually *hydrophilic*. See also Hiemenz and Rajagopalan (1997), who based themselves in part on Overbeek (1952; see above), in their Chapter 13 on "Electrostatic and Polymer-Induced Colloid Stability." Like Overbeek (1952), Hiemenz and Rajagopalan discuss the Schulze (1882, 1883) and Hardy (1900) phenomenon (see their Chapter 13).

Briefly, the Schulze–Hardy phenomenon is as follows: Stable aqueous suspensions ("sols") of (negatively charged) $AsS_2$, Au or AgI particles can be flocculated by the admixture of salts with plurivalent counterions (here *cations*) to the effect that the higher the valency of the added counterions, the less of them is needed to achieve flocculation. The experimental validity of this phenomenon was convincingly demonstrated by Overbeek (1952). The classical explanation of the Schulze–Hardy phenomenon held that the added counterions decreased or even neutralized the $\zeta$-potential of the charged particles so that, with the decrease or the total disappearance of the electrostatic repulsion between the particles, the London–van der Waals attraction prevailed and caused the particles to flocculate and the higher the valency of the added counterions, the less of them were needed to achieve the neutralization of the previously charged particles. All of this was entirely in line with the classical DLVO theory involving the opposing actions of just van der Waals attractions *versus* electrostatic repulsions, but as far as the actual mechanism is concerned, this explanation is not only insufficient, it happens to be completely erroneous. The real explanation of the Schulze–Hardy phenomenon is based on the fact that if one induces a decrease in the $\zeta$-potential of electrically charged hydrophilic particles, these particles' surfaces change from hydrophilic to hydrophobic, which causes them to agglomerate through hydrophobic attraction when they are immersed in water; see Sub-section 3.1, below.

On the other hand the stability of aqueous suspensions of intrinsically electrically neutral hydrophilic particles, such as dextran particles or polyethylene oxide-coated particles, is not affected by pH changes; see the following Sub-section 2.1, below.

## 2.1 Role of attached ionic surfactants or electrically charged polymers in conferring stability to aqueous particle suspensions

There is no fundamental difference between the stabilization of aqueous suspensions of electrically charged inorganic particles and the stabilization of non-charged, initially hydrophobic (e.g., polystyrene latex) particles, by the adsorption or electrostatic or covalent attachment of electrically charged surfactant or of hydrophilic polymer molecules; see, e.g., Hiemenz and Rajagopalan (1997, their Chapter 13).

## 2.2 Non-charged particles or particles of low charge stabilized by non-ionic surfactants or polymers, via "steric stabilization"

### 2.2.1 Definition and origin of "steric stabilization"

The most succinct definition of "steric stabilization" has probably been given by Napper (1983, p. 197) in the title of his Chapter 10: "*Stabilization by Attached Polymer.*"

I do not know when the now exceedingly popular term: "steric stabilization" was first coined, but it was definitely later than 1965. In that year appeared a multi-authored treatise (covering all aspects of what is now generally alluded to as "steric stabilization"), entitled: "Non-Ionic Surfactants" (Schick, 1965, Marcel Dekker, New York), in which at least 10 of the 29 chapters treat different variants of polyethylene oxide, PEO (in biological and medical circles usually called polyethylene glycol, PEG), as well as similar compounds. The work contains a crucial Chapter 19, by Professor R.H. Ottewill (1967), on the "Effect of Non-Ionic Surfactants on the Stability of Dispersions" which defines and explains what is now generally designated as "steric stabilization," without however using that term. Ottewill (1967) begins by mentioning that the flocculation of aqueous dispersions of hydrophobic particles can be prevented by *"protection"* through the admixture of hydrophilic (i.e., water-soluble) polymers, where the term "protection" was first proposed by Zsigmondy (1901); see also Overbeek (1952, p. 316).

On the other hand, in a new multi-authored treatise, also edited by Schick (1987) and also entitled "Non-Ionic Surfactants" and published by Marcel Dekker, New York (1987), "steric stabilization" was defined by Hough and Thompson (1987) in their Chapter 11 on the "Effect of Non-Ionic Surfactants on the Stability of Dispersions." Meanwhile, Napper (1983) already mentioned "steric stabilization" (as early as p. 13) and the subject pervades most of his book. It is interesting to note that the subject "poly(ethylene oxide)" (PEO), or polyoxyethylene is the most quoted item in Napper's (1983) index, as is also the case in Schick's (1965, as well as 1987) "Non-Ionic Surfactants" and PEO is to this day the most utilized suspension-stabilizing polymer. Thus, the term "steric stabilization" must have been coined at some point between 1965 (Schick 1) and 1983 (Napper). Meanwhile Napper (1983) already noted that what is now alluded to as "steric stabilization" was already in use 45 centuries ago in the preparation of "India ink," used in Egypt as well as in China (and presumably also in India, as it is named after that country), around 2500 B.C. (Napper, 1983, p. 18).

### 2.2.2 Reasons why the term "steric stabilization" is neither apt nor explanatory

The term "steric" in "steric stabilization" implies that something that takes up space, or volume, can stabilize particle suspensions in a liquid, which of course is perfectly true but not highly informative. In addition, the term obfuscates the real reason why the particles, thus "sterically" treated, become "stable," i.e., repel one another when immersed in a liquid (usually, water). Many particles (even mineral particles) can spontaneously form stable suspensions in water (e.g., when they are hydrophilic or close to hydrophilic). Examples of this are silica particles and various sorts of clay particles, such as most smectites; see Giese et al. (1996) and Giese and van Oss (2002). What is crucial is that what "steric stabilization" does is to coat hydrophobic particles with something hydrophilic. What is relatively novel is that that "something" can be *non-ionic*, which even nowadays appears to puzzle many workers in this field. Adherents of the classical DLVO theory easily understand that particles with a high enough $\zeta$-potential should be able to repel each other when immersed in water, but the mechanism by which *non-ionic* surfactants or polymers can cause such a mutual repulsion still remains a mystery to them. It seems plausible to many workers in the field that non-ionic polymers such as PEO, when attached to particles, can cause a repulsion by "sterically" waving to one another in a quasi-Brownian manner, but it does not appear to be universally realized that if such "sterically" behaving polymer molecules did *not already repel one another* when immersed in water, such molecules would still be unable to stabilize particles when attached to them but would, instead, be busy cross-linking them.

Now, all these non-ionic "sterically" stabilizing polymers such as PEO, *do repel one another when immersed in water.* In fact, hydrophilic polymer molecules above a given relatively small molecular weight can only dissolve in water *because they mutually repel each other* while immersed in it; see Chapter 7, Sub-section 3.1. Furthermore when polymers are non-ionic (like PEO) they can only be water-soluble by being significantly hydrophilic. In practice this means that such polymer molecules (p), when immersed in water (w) have to have a free energy of interaction with each other, expressed as $\Delta G_{pwp}$, where $\Delta G_{pwp} > 0$ and where the subscript, "p," stands for "polymer"; see also Chapter 5, Sub-section 1.3.

### 2.2.3 Surface properties of "sterically" stabilizing non-ionic polymers and of the polar moieties of non-ionic surfactants

All non-ionic "sterically" stabilizing moieties are hydrophilic, i.e., their $\Delta G_{iwi}$ ($=\Delta G_{pwp}$, see above) $> 0$, which is usually the case when their $\gamma_i^+$ parameter is zero or close to zero and their $\gamma_i^-$ parameter is higher than about 28.5 mJ/m$^2$, which causes them to repel one another in water; see Chapter 5, Sub-section 1.3.1. In other words, all these stabilizing polymers or surfactant moieties have a very elevated term 5 of Eq. (5.1), i.e., they attract water molecules more strongly than they attract each other when immersed in water, and term 5 is positive, so that they *repel* one another as a net result. This property also causes them to be soluble in water; see the previous sub-section, above. Thus, coating or painting hydrophobic particles with any polymer that is soluble in water will bestow the power of aqueous suspension stability on such particles.

It should also be noted that particles that are already hydrophilic, such as silica particles, or various (but not all) clay particles (see Giese et al., 1996), which are to a greater or lesser degree also ionic, nonetheless depend on their hydrophilicity more than on their electrical double layer potential for maintaining suspension stability in water.

(For a linkage between EL and AB properties of particles, see Section 3, below.)

A remaining problem for some readers still may be to understand how non-ionic polymeric or surfactant molecules manage to repel one another, even in the absence of any ionic repulsion. This problem is closely related to the problem inherent in the old, classical DLVO theory, i.e., to try to explain the stability of particles in aqueous suspension, which have been coated with a strictly non-ionic macromolecule, without taking the polar properties of water into account. Long disquisitions on, e.g., brush formation, density of attached polymers, train, loop or tail formation, lengths of attached polymer strands, while extensively treated by many authors (see, e.g., Napper, 1983; Israelachvili, 1991; Hiemenz and Rajagopalan, 1997) do not convincingly clarify the actual mechanism of "steric stabilization." In spite of all these detailed considerations one will look in vain for a quantitative determination of the free energy of repulsion between "sterically stabilizing" polymer strands, when immersed in water, without at the same time looking at the polar properties of water as well as of the stabilizing strands.

Undoubtedly the most important and most quoted and utilized linear polymer, used in "steric stabilization" is polyethylene oxide (PEO), see, e.g., the Subject Indexes of Napper (1983) and Schick (1987) and even Schick (1965) which is the first book on non-ionic surfactants, published well before the term "steric stabilization" was conceived. It is therefore especially useful to determine the surface properties of PEO and from these the free energy of interaction between PEO molecules, when immersed in water. The surface-thermodynamic values for PEO are as follows: $\gamma_i^{LW} = 43.0$ mJ/m$^2$, $\gamma_i^{+} = 0$ and $\gamma_i^{-} = 64$ mJ/m$^2$ (van Oss, 1994, 2006). From these data, using Eq. (5.1), one finds that $\Delta G_{iwi}^{LW} = -7.13$ mJ/m$^2$ and $\Delta G_{iwi}^{AB} = +59.6$ mJ/m$^2$, so that $\Delta G_{iwi}^{IF} = +52.5$ mJ/m$^2$. To obtain the free energies, expressed in units kT (i.e., energy per molecule pair; in this case, of PEO 6000) one also needs to know the contactable surface area ($S_c$) between two PEO molecules, when immersed in water. In the repulsive mode (see Chapter 7, Sub-section 1.1; see also Figure 7.1) this $S_c$ value is 21.2 Å$^2$ (i.e., $21.2 \times 10^{-16}$ cm$^2$) (van Oss, 1994, p. 22; 2006, p. 254). To obtain the $\Delta G_{iwi}^{IF}$ values, given above in mJ/m$^2$, in units kT, one uses:

$$\Delta G_{iwi}(\text{in kT}) = \Delta G_{iwi}(\text{in mJ/m}^2) \times S_c/kT. \tag{8.1}$$

Thus for PEO (6000), $\Delta G_{iwi}^{LW} = -0.37$ kT (at contact, i.e., at $d = d_0$); and $\Delta G_{iwi}^{AB} = +3.12$ kT, each per (crossed) PEO molecule pair (see Figure 7.1B), so that the total interfacial (IF) free energy value per pair of PEO (6000) molecules, immersed in water, amounts to: $\Delta G_{iwi}^{IF} = -0.37 + 3.12 = +2.75$ kT. This is a repulsion which is ample to assure the aqueous solubility of PEO molecules and equally ample to cause a repulsive energy between, e.g., two polystyrene latex particles of about 1 μm in diameter, when each particle has been coated with a sufficient number of PEO molecules to assure that at the relatively close approach of $d \approx 5$ nm all such PEO-coated particles will still repel each other. For instance, at sufficient PEO strands per particle surface such that on average of about 10,000 PEO strands will be directly opposed to 10,000 other PEO strands on a neighboring particle's surface, the repulsive interfacial free energy $\Delta G_{iwi}^{IF} = (\Delta G_{iwi}^{LW} + \Delta G_{iwi}^{AB})$ will amount to about +23,800 kT at "contact" (i.e., at $d = d_0$). At $d = 50$ Å between the distal parts of particle 1 and those of particle 2, the repulsive energy between the two particles is still about +4.9 kT which suffices to prevent them from approaching each other any closer as each decrease in distance to less than 50 Å will ultimately asymptotically increase the interparticle energy of repulsion up to +23,800 kT as $d \rightarrow d_0$, thus assuring particle stability in water. Beyond d = 50 Å, a weak LW attraction takes over which at 80 Å drops to −1.4 kT and at d > 100 Å to an ineffective −0.9 kT and less at greater distances.

Thus, PEO molecules mutually repel one another, when immersed in water, with (in this particular example) a free energy ($\Delta G_{iwi}^{IF}$) = +2.75 kT per molecule pair, which makes them completely soluble in water; see also Chapter 7, Table 7.1. When attached or adsorbed in sufficient numbers onto hydrophobic entities such as polystyrene latex particles, these PEO molecules convey their power of quantitatively measurable mutual Lewis AB interactions-driven repulsion to these particles, which then permits such PEO-coated particles to repel each other when immersed in water, thus stabilizing their aqueous suspension with a measurable free energy of repulsion, solely derived from their PEO coating. This is what really happens when one "*sterically* stabilizes" aqueous suspensions of hydrophobic particles.

### 2.2.4 Residual influence of the underlying hydrophobic particle, when coated with a hydrophobic layer, or brush

It may be useful to consider if, and to what degree, the material itself of a hydrophobic particle can "shine" through a brush of hydrophilic strands. LW, AB and EL properties do this to different degrees and are therefore treated separately, below:

A. LW properties:

Given hydrophobic particles which are coated with a solid thin film of a hydrophilic material, the "screening-out" of LW energies by such a film is rather efficient (see Israelachvili, 1991). Thus with a complete coating with a solid hydrophilic film, the risk of any long-range van der Waals attraction originating from the underlying hydrophobic material of the particles themselves escaping through such a film may be considered negligible. This holds even more for the most hydrophobic underlying materials, such as Teflon, polyethylene or polypropylene which have a rather low Hamaker constant (van Oss, 1994, p. 176; 2006, p. 218). Furthermore, when considering hydrophobic particles with adsorbed or otherwise attached hydrophilic strands that protrude distally to a length of about 100 Å: when, e.g., hydrophobic polystyrene particles are the ones under consideration, then their $\Delta G_{iwi}^{LW}$ at $d = d_0$ amounts to a value of $-6.56\ mJ/m^2$ which in view of London–van der Waals retardation would, at a strand length of 100 Å, only amount to $-0.0016\ mJ/m^2$, which is entirely negligible, as that is the point where the repulsive $\Delta G_{iwi}^{AB}$ of $+59.6\ mJ/m^2$ of, e.g., PEO strands only *begins* (for PEO, see Sub-section 2.2.3, above).

B. AB properties:

Lewis acid–base (AB) interaction energies are propagated only via the surface properties of those immersed molecules or particles situated at the immersed molecule/water interface or particle/water interface, by means of their capacity to orient the vicinal water molecules, which then orient the next layer of adjoining water molecules to a somewhat lesser extent, etc., see Chapter 3, Section 4.

Thus, for a thin solid hydrophilic film which coats a hydrophobic particle, the only AB energies emitted by such particles will emanate from the outer surface of the hydrophilic coating material.

In the case of hydrophobic particles with reasonably densely attached PEO or similar hydrophilic strands, the influence of interstitial hydrophobic patches situated on the hydrophobic particles' surfaces, in between the points of attachment of the hydrophilic strands is largely superseded by the relatively long (≈100 Å) hydrophilic strands stretching into the water. Thus, at a distance of about 100 Å from the particles' surface, the orientation of the water molecules into the bulk water, driven by the hydrophilic strands and projected into the vicinal water, completely predominates over the feeble influence of the distant hydrophobic material of the hidden particles' surface. As discussed above (Sub-section 2.2.3), for R = 1 μm, particles coated with PEO strands, at $d = d_0$ (i.e., at the distal ends of the PEO strands) $\Delta G_{iwi}^{AB}$ equals +23,800 kT.

C. EL properties:

The electrical double layer properties of the naked surfaces of the underlying hydrophobic particles are completely masked when they are solidly coated with a thin hydrophilic layer, so that it will only be the ζ-potential of the hydrophilic coating that counts.

For hydrophobic particles with attached hydrophilic strands stretching outward into the aqueous medium, any electrical double layer energy that can still emanate a repulsive $\Delta G_{iwi}^{EL}$-type energy originating at the naked polymer particle's surface when immersed in water, but sprouting many hydrophilic polymer strands of about 100 Å in length, can only make itself felt to another similarly equipped particle, in water, after having crossed about 100 Å. To interact electrostatically with another such particle the total mutual interaction distance will then be about 200 Å. To estimate the influence of a ζ-potential of about −50 mV, which at contact (i.e., at $d = d_0$), yields a $\Delta G_{iwi}^{EL}$ value of about 3650 kT per pair of particles with a radius, R = 1 μm, which at an interparticle distance of d = 200 Å is reduced to about +6.5 kT (given a rather low ionic strength of the aqueous medium of μ = 0.01; see Chapter 3, Table 3.3). In comparison with a $\Delta G_{iwi}^{AB}$ value at ($d = d_0$) of +23,800 kT (see Sub-section 2.2.3, above) a $\Delta G_{iwi}^{EL}$ contribution of +6.5 kT is exceedingly small. The influence of an underlying ζ-potential emanating from the naked polymer–water interface may therefore be neglected when the hydrophobic particles are coated with brushes comprising hydrophilic strands consisting of PEO, or similar hydrophilic polymers, with strands that are about 100 Å long.

D. Conclusion:

In systems of hydrophobic particles, in aqueous suspension, where the particles are either solidly coated with a thin layer of hydrophilic material, or with a brush of hydrophilic polymer strands, the underlying $\Delta G_{iwi}^{LW}$ and $\Delta G_{iwi}^{EL}$ energies *at the hydrophobic particle–water interface* may be neglected and only the $\Delta G_{iwi}^{LW}$, $\Delta G_{iwi}^{AB}$ and $\Delta G_{iwi}^{EL}$ values *of the hydrophilic coating* or *of the polymer strands' surfaces* need to be taken into account.

# 3. Linkage between the EL Potential and AB Interaction Energies in Water—Importance of AB Interaction Energies for the Stability vs Flocculation Behavior of Aqueous Suspensions of Charged Particles—The Schulze–Hardy Phenomenon Revisited

## 3.1 Mechanism of Schulze–Hardy type flocculation

Since the observations by Schulze (1882, 1883) and Hardy (1900) it has become well-established that when one has an aqueous suspension of electrically charged particles and one causes a decrease in the particles' $\zeta$-potential by the admixture of salts with a plurivalent counterion [i.e., salts like $CaCl_2$, $LaCl_3$, $Th(NO_3)_4$, in the case of *negatively* charged particles], the particles will no longer remain in a stable aqueous suspension but will, instead, flocculate. It was also observed that in order to bring about flocculation of negatively charged inorganic sols, an average of $\approx$74 mM/L salts with a monovalent cation, $\approx$0.93 mM/L salts with a divalent cation, $\approx$0.055 mM/L salts with a trivalent cation and 0.035 mM/L $Th(NO_3)_4$ [averaged from Overbeek (1952, Table 1, p. 307); see also Verwey and Overbeek (1948)]. In other words, to induce flocculation in stable aqueous sols of negatively charged mineral particles (e.g., $As_2S_3$, Au and AgI), one needs about 80 times fewer divalent cations than monovalent ones; about 17 times fewer trivalent cations than divalent ones and 1.7 times fewer tetravalent cations than trivalent ones.

Verwey and Overbeek (1948) and Overbeek (1952) attempted to confirm the correctness of what subsequently became known as the DLVO theory (see Chapter 3), based in part on stability vs. flocculation on Schulze–Hardy systems, compared with measured $\zeta$-potentials and calculated London–van der Waals energies (the latter of which were at the time not yet established with any degree of precision). However, in the 1940's and '50's, given the still prevailing uncertainties concerning the precise values of the Hamaker constants of small particles (see, e.g., Overbeek, 1952, p. 306) and owing to the as yet total ignorance of the polar forces acting on non-polar as well as on polar particles immersed in water (which can be repulsive or attractive and which can change from one into the other as a function of changes in $\zeta$-potential), no clear understanding as yet existed of the roles of the much stronger polar interactions occurring in water which are caused by the always-present hydrophobic attraction due to the hydrogen-bonding free energy of cohesion of the surrounding water molecules, as well as by hydration pressure (i.e., hydrophilic repulsion) which works through the strong hydration energies of hydrophilic surfaces; see also Chapter 3, Section 4 and Chapter 5, Sections 2 and 3. It should therefore not be unduly surprising that one had the impression in the 1940's and up to several decades later, that the classical DLVO approach, involving only London–van der Waals attractions and electrical double layer repulsions should reasonably well describe the experimental results on charged particle stability phenomena in aqueous systems in a satisfactory manner.

It was not until 1991 [Costanzo et al. (1991); see also van Oss, Giese and Norris (1992); van Oss, Giese, Li et al. (1992); van Oss, Giese, Wentzek et al. (1992)] that it became feasible to determine the surface properties of small, non-swelling inorganic particles by contact angle measurements via thin layer wicking, although it was already possible to use direct contact angle measurements on flat layers of particles of swelling clays (i.e., smectites), using hectorite particles (Costanzo et al., 1990, van Oss, Giese and Costanzo, 1990). The method of determination of the $\zeta$-potentials of small particles suspended in an aqueous medium, using microelectrophoresis, had of course been well known since 1932 (see, e.g., Abramson et al., 1942, p. 43), while a simpler microelectrophoresis method, which obviated the operational difficulties and optical complications caused by electroosmotic backflow, was developed somewhat later (van Oss, Fike et al., 1974). The latter microelectrophoresis approach was used by us for measuring the $\zeta$-potentials of inorganic particles in our study on the Schulze–Hardy phenomenon (Wu, 1994; Wu et al., 1994a, 1994b); see below. Once the technique of contact angle determination via thin layer wicking (Costanzo et al., 1990; van Oss, Arnold et al., (1990); van Oss, Giese et al. (1990); van Oss, Good et al. (1990); van Oss, Giese, Li et al., 1992) was also mastered, a quantitative study of the surface energies involved in particle stability in water was started on different Schulze–Hardy particle systems. Negatively charged inorganic particles with radii of the order of 1 μm, were treated with salts comprising plurivalent counterions (in this case, cations), such as $CaCl_2$ and $LaCl_3$. Before and after flocculation of the particles, $\zeta$-potentials were determined via microelectrophoresis (using the electroosmosis-free method developed by van Oss, Fike et al., 1974) and the surface tension components and parameters were determined via thin layer wicking (see above and Chapter 2, Sub-section 1.6.4). The particles used were ground glass particles, calcite particles and smectite clay particles (SWy-1, a montmorillonite clay particle from Southern Wyoming). In Table 8.2 results are shown for calcite particles (Wu, 1994), which could be flocculated by the admixture of (3.9 mM/L) $CaCl_2$, or by (0.47 mM/L) $LaCl_3$. From the measured surface-thermodynamic values of the solid particles (s) (i.e., $\gamma_s^{LW}$, $\gamma_s^+$ and $\gamma_s^-$), the $\Delta G_{sws}^{LW}$ and $\Delta G_{sws}^{AB}$ values were determined (see Chapter 5, Section 1). From the electrophoretic mobilities of the particles, their $\zeta$-potentials and hence their $\psi_0$-potentials (see Chapter 3,Sub-section 5.1) and from these their $\Delta G_{sws}^{EL}$ values (see Table 3.3 for the EL equations) were obtained.

Table 8.2 shows that the stability of aqueous suspensions of calcite particles was for 85.6% assured by AB and for 14.4% by EL repulsions. When flocculated by the admixture of $CaCl_2$, as well as by $LaCl_3$, the particles would have remained in stable suspension if it were not for the newly induced AB attraction and if the classical DLVO energies, i.e., solely van der Waals attraction and electrical double layer repulsion were active in water, the particles would have remained stable, because in both cases the measured EL repulsion still surpassed the LW attraction, by +46 kT (with $CaCl_2$) and by +33 kT (with $LaCl_3$). *In actual fact* the particles were flocculated by $CaCl_2$ as well as by $LaCl_3$, both through strong sign reversals of $\Delta G_{sws}^{AB}$. In other words, the relatively modest decreases in $\Delta G_{sws}^{EL}$ brought about by the admixture of $Ca^{2+}$ as well as by $La^{3+}$, also caused *much stronger changes* in the polar (AB) interaction energies, caused by the fact that the particles switched from quite hydrophilic to severely hydrophobic: $\Delta G_{sws}^{AB}$ changed from *plus* 7,700 kT to *minus* 18,000 kT with the added $Ca^{2+}$ ions and to *minus* 13,000 kT with the added $La^{3+}$ ions (see Table 8.2 and see also Sub-section 3.2, below, on the EL–AB linkage which is one

**Table 8.2** Stability (+ +) vs flocculation (− −) of an aqueous suspensions of calcite particles,[a] "as is" and after addition of small amounts of $CaCl_2$ or $LaCl_3$[b]

| | As is | +3.9 mM $CaCl_2$ | +0.47 mM $LaCl_3$ |
|---|---|---|---|
| Stability | + + | − − | − − |
| $\gamma_s^{LW}$ | 29.1 $mJ/m^2$ | 28.0 $mJ/m^2$ | 30.2 $mJ/m^2$ |
| $\gamma_s^+$ | 0.5 $mJ/m^2$ | 0.3 $mJ/m^2$ | 0 $mJ/m^2$ |
| $\gamma_s^-$ | 31.6 $mJ/m^2$ | 14.2 $mJ/m^2$ | 17.8 $mJ/m^2$ |
| $\Delta G_{SWS}^{LW}$ | −129 kT | −94 kT | −167 kT |
| $\Delta G_{SWS}^{AB}$ | +7700 kT | −18,000 kT | −13,000 kT |
| $\Delta G_{SWS}^{EL}$ | +1300 kT | +140 kT | +200 kT |
| $\Delta G_{SWS}^{TOT}$ | +8871 kT | −17,954 kT | −12,967 kT |
| Electrophoretic mobility (in $\mu m\, s^{-1}\, V^{-1}$ cm) | −2.43 | −0.75 | −0.94 |
| $\zeta$-potential | −34.3 mV | −10.6 mV | −13.3 mV |

*a* Calcite particles: 0.5% (w/w) aqueous suspension; ionic strength: $\mu = 0.015$; pH $\approx 7.7 \pm 0.2$; average particle radius, R = $10^{-4}$ cm.
*b* Data from Wu (1994), with permission.

of the underlying causes of the AB shift from hydrophilicity to hydrophobicity as a result of a decrease in $\zeta$-potential).

Wu (1994) observed that the decrease in the particles' $\zeta$-potential necessary to cause flocculation would not have sufficed to decrease the $\Delta G_{sws}^{EL}$ value of the system enough to cause the calcite particles to flocculate according to the old, classical DLVO theory, because the residual positive value of $\Delta G_{sws}^{EL}$ still denoted a repulsion that was greater than the negative (i.e., attractive) value of $\Delta G_{sws}^{LW}$ after treatment with $CaCl_2$ as well as with $LaCl_3$ (see also Wu et al., 1994a, 1994b). At the same time Wu (1994) and Wu et al. (1994a, 1994b) observed that the primary underlying mechanism of the destabilization of charged inorganic particles caused by the Schulze–Hardy type of admixture to the aqueous particle suspension of salts endowed with plurivalent counterions with respect to the charged particles, was the concomitant strong decrease in the value of the $\gamma_s^-$ parameter of the surfaces of the charged particles (s). Decreases in the $\gamma_s^-$ value cause charged particles to become less hydrophilic, or to switch from hydrophilic to downright hydrophobic. With the calcite particles the original value for $\gamma_s^-$ was 31.6 $mJ/m^2$, when the particles were still in stable aqueous suspension. Upon the admixture of 3.9 mM/L $CaCl_2$ that value decreased from 31.6 to 14.2 $mJ/m^2$ and after adding 0.47 mM/L $LaCl_3$ $\gamma_s^-$ decreased from 31.6 to 17.8 mM/L. (In all cases, treated or untreated, the $\gamma_s^+$ values for calcite were negligibly small; see Wu, 1994; Wu et al., 1994a, 1994b.)

The classical hypothesis based on the old DLVO theory, proposed in explanation of the Schulze (1882, 1883) and Hardy (1900) phenomenon of flocculation of stable aqueous inorganic particle suspensions by the addition of inorganic salts with plurivalent counterions was that the admixture of such counterions decreases the electrostatic potential of the particles to the point where the London–van der Waals attraction takes over and causes the particles to flocculate (Verwey and Overbeek, 1948; Overbeek, 1952). However, Wu (1994) and Wu et al. (1994a, 1994b) demonstrated to the contrary (see Table 8.2) that the predominant mechanism of destabilization of aqueous suspensions of (in this case negatively) charged particles by means of the admixture of salts with plurivalent cations, is the marked decrease in the electron-donicity (i.e., the $\gamma_s^-$ value) of the particles, caused by the addition of strong electron-acceptors in the guise of the added plurivalent cations, thus neutralizing most of the particles' electron donicity, causing them to reverse their polar surface properties from hydrophilic to hydrophobic, which then exposes them to a strong mutual hydrophobic attraction in water, i.e., to flocculation. Thus, the observed decrease in the particles' $\zeta$-potential is simply a small secondary effect and not the direct cause of the hydrophobization which is directly caused by the admixture of the plurivalent cations.

It should be noted that the plurivalent cation-induced flocculation of negatively charged inorganic particles is reversible through repeptization by the admixture of $Na_2EDTA$ or $Na_3EDTA$ or $(NaPO_3)_6$, which are complexing agents for plurivalent inorganic cations and thus are capable of removing these cations. Upon repeptization by means of any one of these complexing compounds, the particles also regain their original high electron-donicity and with it, their pronounced hydrophilicity. On the other hand, attempts to remove the $Ca^{2+}$ or $La^{3+}$ ions by thorough washing with water and thus to induce repeptization, proved to be completely ineffective (Wu, 1994; Wu et al., 1994a, 1994b).

## 3.2 Linkage between changes in $\zeta$-potential and especially, changes in the electron-donicity of polar surfaces, when immersed in water

Our study on the destabilization of mineral particle suspensions in water through the admixture of plurivalent counterions (Wu, 1994; Wu et al., 1994a, 1994b; see also the preceding Sub-section 3.1, above) led to the discovery of a one-way linkage between electrical double layer potentials (i.e., $\zeta$-potentials) and polar (Lewis acid–base: AB) interaction energies acting on charged polar particles while immersed in water. Non-charged polar particles or macromolecules such as dextran or polyethylene oxide (PEO) on the other hand, are not influenced by the addition of plurivalent ion-containing salts, nor by pH changes. The addition of plurivalent counterions, or the application of pH changes, can only change the degree of hydrophilicity or hydrophobicity of *electrically charged surfaces*, or *potentially electrically charged surfaces*, such as amphoteric surfaces, whereas uncharged or non-amphoteric surfaces are impervious to such actions. Thus, measures used to change the $\zeta$-potential of charged polar surfaces can only change the sign and value of $\Delta G_{sws}^{AB}$ when the solid polar surface (s) has an electric charge. At an amphoteric surface's minimum (or zero) electric charge, its $\Delta G_{sws}^{AB}$ tends to be negative (i.e. hydrophobic) and its degree of hydrophobicity is at its highest value at its isoelectric pH. For instance in the field of protein purification one prefers to precipitate proteins at their isoelectric pH and after washing the precipitate one then redissolves them most easily at a different (usually higher) pH. The addition of salts with plurivalent cations (for negatively charged proteins, usually the addition of $CaCl_2$) has the same effect as a decrease in the ambient pH of their solution, both of which render such proteins more hydrophobic (and also decrease their aqueous solubility). This also applies to negatively charged phospholipids, see, e.g., van Oss et al. (1988); van Oss (1994, p. 218; 2006, p. 251).

To summarize: electrically charged surfaces or potentially electrically charged (amphoteric) surfaces can be made to switch from hydrophilic to hydrophobic by causing a decrease in their $\zeta$-potential, *or* by directly neutralizing their electron-donicity with electron-accepting cations (which concomitantly also decreases their $\zeta$-potential, as a side-effect). As also mentioned in Sub-section 3.1, above, the hydrophobizing action of added electron-acceptors in the guise of plurivalent cations can be reversed by the removal of these cations with complexing agents such as EDTA.

On the other hand, admixture of non-charged hydrophilic polymers with a high $\gamma_s^-$ value such as PEO (with $\gamma_s^- = 64$ mJ/m$^2$) or dextran (with $\gamma_s^- = 55$ mJ/m$^2$), do *not* influence the $\zeta$-potential of charged particles in aqueous suspensions. *This is in contradiction to conclusions by Brooks and Seaman* (1973) *and and by Brooks* (1973), who believed they had observed just such an influence. They had, however, misinterpreted their own results. When they calculated the $\zeta$-potential from their measured electrophoretic mobilities of negatively charged red cells after the addition of dextran, they did not account for the increased viscosity of their aqueous red cell suspension caused by the added dextran; see Arnold et al. (1988) and see also van Oss (1994, p. 64; 2006, p. 67). In reality, adding dextran to a suspension of charged particles does not change their $\zeta$-potential at all; it only slows down their electrophoretic mobility because the added dextran increases the viscosity of the liquid medium; see Chapter 3, Eq. (3.8). Meanwhile, there is one thing which added dextran does to red cells: it crenates them, i.e., it changes their surfaces from smooth to bumpy, which can facilitate the ease by which red cells can be cross-linked, which causes them to flocculate by antibody-induced hemagglutination; see van Oss, 1994; 2006; see also Sub-section 4.2, below. This change in the surface structure of red cells through the admixture of dextran, dissolved in the common aqueous medium is due to the very high osmotic pressures hydrophilic, neutral polymers can exert, which can amount to hundreds of atmospheres at perfectly attainable polymer concentrations; see Chapter 5, Section 4.

Finally, another form of EL/AB linkage can be observed in non-aqueous polar organic liquid media. For instance, the degree of electron-*donicity* of a polymer or a particle, immersed in a polar, electron-*accepting* organic liquid, can be estimated from its electrophoretic mobility in that liquid. The same can be done with an electron-*accepting* polymer or particle, immersed in an electron-*donating* polar organic liquid. (Fowkes et al., 1982; Labib and Williams (1984, 1986, 1987); see also van Oss, 1994, p. 219; 2006, p. 252). Unfortunately, most likely as a consequence of the complications and even dangers of carrying out electrophoresis (which develops considerable amount of heat) in often flammable organic liquids, it does not appear that this methodology has gained many followers.

See also Chapter 3, Section 7 for other related observations on the EL/AB linkage.

# 4. Destabilization of Aqueous Particle Suspensions by Cross-Linking

Cross-linking and subsequent flocculation of aqueous suspensions of particles by the addition of soluble linear polymers can occur as a side effect, despite the fact that the intention may only have been to stabilize the particles. This usually unwanted side effect occurs most readily at low polymer concentrations, see Hiemenz and Rajagopalan (1997, p. 605).

On the other hand, flocculation of particles or cells in aqueous suspensions is frequently done intentionally, for diagnostic purposes, in hematology and blood transfusion. One of the oldest uses of this approach is used for blood group typing in blood banking, with a view to determining the blood groups of blood donors as well as of recipients. Another, also rather well-established application is the latex fixation test, in particular for the determination of the presence of rheumatoid factor in the blood serum of patients with rheumatoid arthritis.

## 4.1 Cross-linking of latex particles for diagnostic purposes—The latex fixation test

The latex fixation test for the detection of rheumatoid factor typically uses monodispersed, approximately micron sized, intrinsically hydrophobic polystyrene latex particles. Such latex particles are stabilized to form a stable suspension in water by the admixture of, e.g., sodium dodecyl sulfate and are activated for use by the addition of normal human polyclonal serum immunoglobulin-G (IgG), which readily adsorbs onto the latex particles and in so doing slightly alters its structure in the same manner by which the structure of class IgG antibodies are slightly altered[1]) when combining with their corresponding antigen. In that form the particles with adsorbed IgG now binds to rheumatoid factor which is an IgM class immunoglobulin with antibody-specific affinity for the type of reacted human IgG that has been adsorbed onto the latex particles. Rheumatoid factor IgM is found in the serum of patients with rheumatoid arthritis. IgG is a divalent immunoglobulin with a molecular mass of about 150 kD. IgM is a decavalent immunoglobulin (Edberg et al., 1971) with a molecular mass of about 900 kD. The large starfish-shaped IgM molecules, with a diameter of about 30 nm can readily cross-link these latex particles with adsorbed IgG molecules on its surface, thus causing the particles to start flocculating within seconds after the addition of a serum sample containing rheumatoid factor. The flocculation is easily observed visually, as the aspect of the suspension changes from smooth milky white, to coarse and grainy, whether in a small test

[1] It has been assumed, to serve as an explanation, that rheumatoid factor IgM molecules bind to the normal IgG molecules after these had been adsorbed onto the latex particles, which putatively causes the structure of the IgG molecules to become slightly altered upon adsorption onto the polystyrene surface of the latex particles, because IgM does not normally adsorb onto single freely dissolved IgG molecules.

Another and simpler explanation may well be that IgM does not normally bind to single separate IgG molecules while they are both dissolved in blood serum. However IgM may be able to bind to several IgG molecules at a time when the latter find themselves bound to a polystyrene surface in close vicinity to one another, so that one IgM molecule can bind to three or four IgG molecules at a time, which considerably enhances the total binding energy of the IgM–$(IgG)_n$ complex thus formed (where $n = 3$ or 4); see also Chapter 14, Sub-section 2.1.6.

tube, or deposited on a glass slide. The transition of completely stable to completely flocculated usually takes about one or two minutes. By using serial dilutions of the IgM solution (patient's serum) to add to aliquots of a stable IgG/latex suspension, a titer can be obtained of the patient's rheumatoid factor.

Similar latex fixation tests have been used as pregnancy tests, e.g., using patients' human urinary gonadothropic hormone (HGH), which inhibits the agglutination (flocculation) of suspensions of latex particles coated with HGH, when admixed with an anti-HGH antiserum which had first been mixed with the patient's urine sample.

## 4.2 Cross-linking of human red blood cells with antibodies to cause flocculation (hemagglutination) for blood group determinations

There are literally dozens of human red cell-linked blood group categories. However, in most cases it suffices to determine blood groups of only two different systems, i.e., the ABO blood groups, distinguishing among donors and/or recipients of red cells belonging to these blood groups (comprising A, B AB and O blood groups) and the Rh blood groups (comprising $Rh_0$, rh′, rh,″ hr′ and hr″, also named D, C, E, c and e, respectively). The most important one of the latter system is the $Rh_0$ blood group, where it is important to know if a blood donor or recipient is $Rh_0$ (often just called Rh) positive or negative. In normal blood transfusion practice the two systems are grouped and described together by combining the ABO and the Rh system. For instance if somebody's red cells are called: "*A positive*" that means that the person in question has *blood group A and is Rh positive*. The tests for these two blood group systems however are somewhat different.

The antibodies against red cells of the ABO system are usually of the IgM isotype, mainly because the antigenic determinants (epitopes) of the ABO systems are carbohydrates, which preferably elicit IgM class antibodies. It should be noted that type O red cells lack A, B, or AB epitopes; in other words, for ABO typing one only needs anti-A and anti-B antibodies; when negative with both, the blood group is O and when positive with both, the blood group is AB. The order of occurrence of the ABO blood groups among people of European descent is: O, A, B and AB, where O and A are the most common blood groups and AB the rarest and where the incidences of O and A are fairly similar and where O is usually the most common, although A can be more common than O in some populations (Mohn, 1979). One advantage of testing for the ABO system blood groups is the fact that the test antibodies usually are of the IgM isotype, with its 10 valencies and its large diameter ($\approx$30 nm; see above), so that the cross-linking of two or more red cells readily occurs. One determines that red cells are, for instance, blood group A when one observes a visible flocculation ("hemagglutination") soon after the addition of anti-A antibodies. Positive hemagglutination is quite visible because one sees the transition from a shiny red, homogeneous-looking drop of blood, to a coarse agglomeration of grainy particles. This is further helped by the fact that under the influence of a pinching action by the IgM-class antibodies, the red cell surface changes from smooth to crenated, or bumpy, as seen via scanning electron microscopy (van Oss and Mohn, 1970). The bumps have a much smaller radius of curvature than the red cells themselves, which aids in decreasing the normally occurring acid–base-driven repulsion between cells and antibodies and thus enhances the opportunity for cross-linking the cells with antibodies (van Oss, 1984a). Added dextran can help red cells become crenated (i.e., "bumpy"), which gives them protuberances of a much smaller radius of curvature than the red cells major radii, which in turn helps antibody molecules to overcome the Lewis acid–base-driven repulsion between red cells and immunoglobulins and thus favors hemagglutination by antibodies; see van Oss, 1984a; van Oss and Mohn, 1970; see also Sub-section 3.2, above. Another reason why hemagglutinating type A or B red cells' is relatively easy lies in the fact that these cells' surfaces (like those of most other blood cells) are submicroscopically not smooth but equipped with protruding glycoprotein mini-fibers which are about 10 nm long. Hemagglutination of blood group A or B erythrocytes is greatly facilitated by the fact the A or B glucide-type epitopes are placed at the extreme distal (outside) ends of these mini-fibers, which together form the glycocalyx of these cells.

With $Rh_0$ (D) positive cells, hemagglutination is more difficult. To begin with, anti-$Rh_0$ antibodies tend to be of the IgG isotype. Now, IgG is only divalent and has a maximum "reach" between its two valencies of only about 6 or 7 nm, whilst the total distance between its antibody-active sites (paratopes) and its Fc-tail is just 10 nm. In addition, the $Rh_0$ (D) epitope is *situated on the red cell membrane*, i.e., about 10 nm below the distal ends of the glycocalyx's fibers. Thus, the IgG anti-$Rh_0$ (D) paratopes, when binding to the $Rh_0$ (D) epitopes on the red cell membranes, disappear almost completely between the red cells' glycocalyx strands, where they tend to bind with both valencies to neighboring epitopes situated on one single cell's cell membrane, well below the tops of the glycocalyx strands, thus thwarting the possibility of cross-linking the cells by such an antibody. In addition, anti-$Rh_0$ (D) binding to the cells' $Rh_0$ (D) epitopes by IgG isotype antibodies, does not crenate the red cells (van Oss, 1984a; van Oss et al., 1979), which makes cross-linking the red cells by IgG-class antibodies even more difficult; see van Oss and Mohn (1970).

One of the better ways to hemagglutinate $Rh_0$ (D)-positive red cells is to follow a two-step procedure. One first binds IgG-class anti-$Rh_0$ (D) antibodies to the red cells that are to be tested; these IgG-class antibodies bind with both valencies to a single red cell's surface (below the glycocalyx, but with their Fc tails just barely distally protruding beyond its strands). Then one adds (e.g., rabbit) antibodies to IgG (which themselves are also of the IgG isotype), to cross-link these red cells which have already been treated with anti-$Rh_0$ (D) antibodies, thus causing the cells to flocculate, i.e., to show a positive hemagglutination result for the $Rh_0$ (D) blood group when the tested red cells are indeed $Rh_0$ (D) positive.

SECTION C

# Physical and Physicochemical Properties of Water

CHAPTER NINE

# Cluster Formation in Liquid Water

## Contents



## 1. Size of Water Molecule Clusters

### 1.1 Measurement of the cluster size of water via its solubility in organic solvents

The solubility equation (van Oss and Good, 1996; see also Chapter 7, Eq. (7.2)):

$$\Delta G_{iwi}^{IF} \cdot S_{c(i)} = kT \cdot \ln s_{(i)} \tag{9.1}$$

[where $\Delta G_{iwi}^{IF}$ is the interfacial free energy, in mJ/m$^2$, of the interaction between two molecules (i), immersed in water (w), at "contact" (i.e., at $d = d_0$); $S_{c(i)}$ is the contactable surface area between two molecules (i), also at "contact" and $s_{(i)}$ is the aqueous solubility of molecules (i), expressed in mole fractions] will allow the determination of the contactable surface area per molecule pair of (i), if one knows the value of $\Delta G_{iwi}^{IF}$ as well as of $s_{(i)}$. In addition, because $S_{c(i)}$ is necessarily a function of the molecular weight of (i), the determination of $S_{c(i)}$, for water, will not only furnish a confirmation of the existence of clusters of water molecules in liquid water, but it can also yield a measure of the cluster *size* of water at a given temperature

*Interface Science and Technology*, Volume 16
ISSN 1573-4285, doi: 10.1016/S1573-4285(08)00209-3

(e.g., at 20 °C). Now, as:

$$\Delta G_{iwi}^{IF} \equiv -2\gamma_{iw} \quad (9.2)$$

one may also write Eq. (9.2) as:

$$\Delta G_{wiw}^{IF} \equiv -2\gamma_{iw} \quad (9.2A)$$

so that another version of Eq. (9.1) then becomes:

$$\Delta G_{wiw}^{IF} \cdot S_{c(w)} = kT \cdot \ln s_{(w)}, \quad (9.1A)$$

where $s_{(w)}$ is the *solubility of water in an organic liquid* (i), which we may call (o), so that:

$$\Delta G_{wow}^{IF} \cdot S_{c(w)} = kT \cdot \ln s_{(w)}. \quad (9.1B)$$

Equation (9.1B) shows that it should be possible to determine the contactable surface area, $S_{c(w)}$ of water molecules, immersed in an organic liquid (o), such as: chloroform, benzene, toluene, xylene, or cyclohexane. Applying Eq. (9.1B), van Oss, Giese and Good (2002) showed, using the solubilities of water in these organic liquids, that the $S_{c(w)}$ for water at 20 °C, when dissolved in the aforementioned organic liquids is:

$$35.78\ \text{Å}^2\ [\pm 1.35\ \text{Å (S.D.)}].$$

Comparing this value with the contactable surface area of monomeric water molecules (when participating in a cluster) of approximately 8 Å$^2$ [see Rao (1972) and Eisenberg and Kauzmann (1969), as discussed in van Oss, Giese and Good (2002)], the cluster's $S_{c(w)}$ value of ≈35.8 Å$^2$ would indicate that at 20 °C, water molecules form clusters comprising, on average, 4.5 $H_2O$ [±0.16 (S.D.)] *molecules per cluster.*

The hydrogen bonds in these clusters break and reform within time-spans of the order of a picosecond ($=10^{-12}$ s) according to Luzar and Chandler (1996), while according to our results an average cluster size (n) of water molecules at 20 °C of n = 4.5 prevails at any given moment. This observation is most compatible with a system of an exceedingly quickly fluctuating cluster formation alternating with an equally fast cluster break-down, caused by an extremely rapid hydrogen-bond association/dissociation cycle, giving rise to an average water cluster size of 4.5 $H_2O$ molecules per cluster, at room temperature.

## 1.2 Variability as a function of temperature (T) of the cluster size as well as of the viscosity of water

There are significant indications that the cluster size of water diminishes significantly upon heating. One indication is the strong decrease in the viscosity of water when increasing T. When heating liquid water by 80°, i.e., from 20 to 100 °C, its viscosity decreases three and a half times, i.e., from $\eta = 1.00$ cP at 20 °C to $\eta = 0.284$ cP at 100 °C. Such strong decreases in viscosity with an increase in T are mainly found in polar, self-associating liquids and among these water is an especially striking example. Another indication is the increased electron-accepticity (i.e., Lewis acidity) of liquid water with an increase in T; see the following Sub-section 1.3, below.

### 1.3 When water cluster size decreases with an increase in T, its Lewis acidity increases and its Lewis alkalinity decreases

Upon heating water from 20 to 38 °C, $\gamma_w^{LW}$ only decreases from 21.8 to about 21.0 mJ/m$^2$ and $\gamma_w^{AB}$ decreases from 51.0 to about 49.9 mJ/m$^2$. Now, whilst the $\gamma_w^{AB}$ values, which are equal to the *product*: $2\sqrt{(\gamma_w^+ \cdot \gamma_w^-)}$, vary little with an increase in T from 20 to 38 °C, the *ratio*: $\gamma_w^+/\gamma_w^-$, which has been established at unity at 20 °C (see Chapter 2, Section 2), *increases by* 75% to: $\gamma_w^+/\gamma_w^- = 1.75$ at 38 °C (van Oss, 1994, pp. 299–301; 2006, pp. 94–96). This therefore changes the values at 20 °C from $\gamma_w^+ = \gamma_w^- = 25.5$ mJ/m$^2$, to $\gamma_w^+ = 32.4$ mJ/m$^2$ and $\gamma_w^- = 18.5$ mJ/m$^2$ at 38 °C. The large increase in Lewis acidity with an increase in T is in agreement with the increase in the number of non-hydrogen-bonded H atoms of water at higher temperatures noted by Nemethy and Scheraga (1962) and by Eisenberg and Kauzmann (1969, pp. 164–165).

It would seem plausible via an extrapolation of known data, that when one approaches the boiling point of water (at ambient atmospheric pressure), the cluster size of liquid water will tend completely toward the monomeric form (van Oss, 2006, p. 98), thus facilitating the escape of single water molecules as steam upon continued heating.

## 2. Implications of the Increased Lewis Acidity of Water Following Increases in T

### 2.1 Consequences for the aqueous solubility of solutes and for the stability of aqueous suspensions as a function of T

#### 2.1.1 Influence of heating on *hydrophobic* solutes or particles, immersed in water

Reiterating Eq. (5.1):

$$\begin{aligned}\Delta G_{iwi}^{IF} &= -2\gamma_{iw} \\ &= \underbrace{2(\sqrt{\gamma_i^{LW}} - \sqrt{\gamma_w^{LW}})^2}_{1} \underbrace{-4\sqrt{(\gamma_i^+\gamma_i^-)}}_{2} \underbrace{-4\sqrt{(\gamma_w^+\gamma_w^-)}}_{3} \underbrace{+4\sqrt{(\gamma_i^+\gamma_w^+)}}_{4} \\ &\quad \underbrace{+ 4\sqrt{(\gamma_i^-\gamma_w^+)}}_{5}\end{aligned} \tag{9.3}$$

one will note that for completely *apolar* solutes or particles (i), terms 2, 4 and 5 are zero, leaving just terms 1 and 3:

$$\Delta G_{iwi}^{IF} = 2(\sqrt{\gamma_i^{LW}} - \sqrt{\gamma_w^{LW}})^2 - 4\sqrt{(\gamma_w^+\gamma_w^-)}. \tag{9.4}$$

Upon heating $\gamma_w^+$ increases and $\gamma_w^-$ decreases, but their *product* (term 3 of Eq. (9.3) only decreases very slightly, so that the (small) aqueous solubility of completely hydrophobic solutes or the (tiny) degree of stability of completely hydrophobic particles when immersed in water, will only slightly increase when the absolute value of term 3 (of Eqs. (5.1) or (9.3)) decreases through an increase in T.

### 2.1.2 Influence of heating on *hydrophilic* solutes or particles immersed in water

For hydrophilic solutes or particles, immersed in water (which hydrophilic solutes or particles invariably have an elevated $\gamma_i^-$ value), the increase in $\gamma_w^+$ with an increase in T very significantly augments the value of term 5 in Eq. (9.3); see above. An increase in the repulsive hydration term 5 decreases the negative value of $\Delta G_{iwi}$, or changes $\Delta G_{iwi}^{IF}$ from negative to positive, or even causes $\Delta G_{iwi}^{IF}$ (when already positive) to become even more positive. Thus an increase in the positive value of term 5 (Eq. (9.3)) strongly enhances the repulsion between solutes or particles (i), when immersed in water, thus promoting solute solubility, as well as particle stability, in water.

The mechanism discussed above is unrelated to the *decrease* in aqueous solubility of certain solutes such as polyethylene oxide (PEO) when reaching their $\theta$ temperature. The $\theta$ temperature is the point where the effect of a further rise in T causes the secondary structure of the polymer to change by exposing its originally hidden hydrocarbon backbone to the polymer–water interface, thus causing the exceedingly soluble PEO to revert to complete aqueous insolubility upon further heating. In the case of normally quite soluble biopolymers such as many proteins, heating of such proteins dissolved in water may give rise to a sudden insolubilization upon a further increase in T, which is often alluded to as heat-induced "denaturation." There is however one (urinary) protein, called "Bence Jones protein," which occasionally occurs among patients with multiple myeloma. This protein which consists of a dimer comprising two immunoglobulin light chains (Mw $\approx$ 40,000) becomes insoluble when a solution is heated to about 60 °C, but redissolves when further heated to between 95 and 100 °C (Rose, 1979; see also van Oss, 1994; pp. 183–184; 2006, p. 220).

Other very soluble hydrophilic biopolymers such as various polysaccharides (e.g., dextran) do not undergo any change in aqueous solubility with an increase in T.

## 2.2 Consequences for the attachment or detachment among two different solutes and/or solids, immersed in water, as a function of T

We are now referring to Eq. (6.1), which plays a similar role as Eq. (5.1) (see also Eq. (9.3), above) but which treats, instead, the free energies between two *different* entities, 1 and 2, immersed in water. The equivalents of term 5 of Eq. (5.1) (see also Eq. (9.3) above) which one may also entitle the principal hydration term, whose positive value increases when $\gamma_w^+$ increases with an increase in T, are the positive $\sqrt{(\gamma_w^+\gamma_1^-)}$ and $\sqrt{(\gamma_w^+\gamma_2^-)}$ terms of Eq. (6.1), whose positive values also increase with an increase in T. This then gives rise to an increased repulsion between polar entities 1 and 2, when immersed in water.

One example, well known to most readers, is the increased ease of detachment of "dirt" from soiled surfaces when washed with warm rather than with cold water (even in the absence of soap).

Another example is the gel formation by linear polymers (see Chapter 7, Sub-section 3.5), at lower T, and the redissolution of the polymers upon an increase in T, where in certain cases the polymer moieties which bind together upon cooling of the aqueous polymer solution have different chemical compositions that allow them to bind together in a complementary fashion. In all such cases Eq. (6.1) will be more applicable than Eq. (5.1), because it allows treating the binding between two *different* sub-entities.

## 3. Influence of Cluster Formation in Liquid Water on the Action at a Distance Exerted by Polar Surfaces when Immersed in Water

### 3.1 Connection between cluster size and the decay length of water

The cluster formation of water molecules in liquid water at temperatures around 20 °C is of considerable importance in the estimation of Lewis acid–base energies of interaction ($\Delta G_{iwi}^{AB}$) at a distance (d) between molecules, macromolecules, particles and cells (i) when immersed in water (w). This is because it is the characteristic length, or "decay length" ($\lambda$) of water, which is taken to be equivalent with the radius of gyration ($R_g$) of the water molecules and which is the crucial constant, at a given temperature, that determines the distance to which a significant free energy ($\Delta G_{iwi}^{AB}$) of attraction or repulsion can be exerted; see Chapter 3, Table 3.2 for the role of $\lambda$ in the exponential decay between polar entities, immersed in water.

If liquid water were simply monomolecular (Mw = 18) $H_2O$, water would have a decay length, $\lambda \approx 2$ Å (Chan et al., 1979; Parsegian et al., 1979). However, at $\lambda \approx 2$ Å, Lewis acid–base (AB) interaction energies would decrease so steeply with distance, d, that even before reaching a distance of only 20 Å, the AB repulsion is already too feeble to overcome the Lifshitz–van der Waals (LW) attraction between two human red cells under *in vivo* conditions, which, given the cell-shape fluctuations that normally manifest themselves in the glycocalyx of circulating erythrocytes, would inexorably lead to the flocculation of these cells; see Table 9.1. Fortunately however, thanks to the existence of water clusters of $R_g = \lambda \approx 6$ Å (see Chapter 3, Sub-section 4.1 and see also van Oss, 1990; 1994, pp. 318–327; 2006, pp. 336–339) corresponding to the occurrence of clusters of about 4.5 water molecules per cluster at 20 °C (see Sub-section 1.1, above), the repulsion reach of AB interaction energies in water extends significantly farther (see Table 8.1) at a value for $\lambda$ of about 6 Å. However a value for $\lambda$ of 10 Å (see Israelachvili and Pashley, 1984) is clearly too high as it would correspond to a cluster size for water molecules at room temperature that is significantly greater than n = 4.5 $H_2O$ molecules per cluster.

One can estimate the value for $R_g$ in the case of a significantly asymmetrical molecule or cluster of molecules, of length, L (Hiemenz and Rajagopalan, 1997), as:

$$R_g = L/\sqrt{12} \qquad (9.5)$$

**Table 9.1** The same as Table 8.1, except for the $\Delta G_{iwi}^{AB}$ values at $d > d_0$, where $\lambda = 10$ Å has been replaced by $\lambda = 2$ Å which would be the value if water did not form clusters. Free energy, i.e., $\Delta G_{iwi}$, in kT units, between human erythrocytes, immersed in water, at ≈35 °C; ionic strength, $\mu = 0.145$ and pH 7.2, as a function of distance d between cells for LW, AB and EL interactions[f]

| $\Delta G_{iwi}$ (in kT) | Distance (d) | | | | | |
|---|---|---|---|---|---|---|
| | $d = d_0$ (=1.57 Å) | 20 Å | 50 Å | 60 Å | 70 Å | 80 Å |
| $\Delta G_{iwi}^{LW}$ | $-1979^a$ | $-155.4$ | $-62.1$ | $-1.36^d$ | $-0.99^d$ | $-0.76^d$ |
| $\Delta G_{iwi}^{AB}$ | $+206{,}800^b$ | $+20.6$ | $+0.0063$ | $+4.7 \times 10^{-8}$ | $+2.86 \times 10^{-10}$ | 0 |
| $\Delta G_{iwi}^{EL}$ | $+1368^c$ | $+107.9$ | $+2.6$ | $+0.76$ | $+0.22$ | $+0.06$ |
| $\Delta G_{iwi}^{TOT\,e}$ | $+206{,}189$ | $-26.9$ | $-59.5$ | $-0.60$ | $-0.77$ | $-0.70$ |

*a* From data in van Oss (1994, 2006), adjusted to 35 °C; see equation for two equal spheres given in Table 3.1.

*b* As in [b] (see above), with equation for two equal spheres in Table 3.2.

*c* From electrophoretic mobilities of human erythrocytes, at 35 °C, $\mu = 0.145$ and pH 7.2 one obtains $\zeta = -10.9$ mV $\zeta = -10.87$ mV and $\Psi_0 = -15.8$ mV (see Chapter 3, Section 5). The $\Delta G_{iwi}^{EL}$ values at $d > d_0$ were calculated using Table 3.3, for the case of two equal spheres.

*d* $\Delta G_{iwi}^{LW}$ is much reduced at $d > 50$ Å because at $d > 50$ Å full van der Waals–London retardation sets in; see Chapter 3, Sub-section 3.2.

*e* It will be seen that if water did not form clusters, i.e., for $\Delta G_{iwi}^{AB}$, $\lambda = 2$ Å, the deciding influence of $\Delta G_{iwi}^{AB}$ would have disappeared by $d = 20$ Å and from d = slightly less than 20 Å, erythrocyte instability would prevail.

*f* In view of the somewhat unwieldy doughnut-like shape of human erythrocytes, their averaged radius has been estimated at $5 \times 10^{-4}$ cm.

(see also van Oss, 1994, pp. 69–70; 2006, p. 74). The length, L of a chain of about 4.5 water molecules (averaged between chains of 4 and chains of 5 water molecules) would be: $L = R_g \cdot \sqrt{12} = 2.078$ nm which, given the total contactable surface area of a cluster of 0.358 nm and an $R_g = \lambda$ value of 0.6 nm (see Sub-section 1.1, above), yields an average cluster width of 0.172 nm.

For a somewhat more compact but still asymmetrical water cluster, its length, L′, may be expressed (Hiemenz and Rajagopalan, 1997), as:

$$R_g = L'/\sqrt{6} \tag{9.6}$$

where the length, L′, of a cluster of about 4.5 water molecules would be equal to: $L' = R_g \cdot \sqrt{6} = 1.47$ nm which, given the total contactable surface area of the cluster as 0.358 nm$^2$ (see Sub-section 1.1, above), yields an average water cluster width of 0.244 nm.

Given the considerable influence of the water cluster size on the extent to which polar interactions between molecules, macromolecules, particles or cells, when immersed in water, are transmitted at some non-negligible distance, it is reasonable to assume that the molecular shape of water clusters is asymmetrical, but it is not possible at this time to decide between an asymmetry ratio of about $2.078/0.172 = 12$ (Eq. (9.5)) or a ratio of about $1.47/0.244 = 6$ (Eq. (9.6)). However, in view of the fact that for the contactable surface area of water clusters at room temperature of about 0.36 nm$^2$ (see Sub-section 1.1, above), one has the same $R_g$ (and thus $\lambda$)

value of 0.6 nm in both cases, one will find the same $\Delta G_{iwi}^{AB}$ *versus* distance values for either assumption (see Table 8.1).

## 3.2 The influence of cluster formation in liquid water on the XDLVO approach pertaining to the stability of aqueous suspensions of human blood cells

Table 9.1 shows that if no water clusters were formed in liquid water and $R_g$ and therefore $\lambda$ would not exceed 2 Å, the AB interaction energies at a distance would be so reduced that under *in vivo* conditions in human peripheral blood, blood cell stability (and even serum protein solubility) would be impaired. Table 9.1 also shows that at d $\approx$ 19 Å (i.e., just before reaching 20 Å) the initially strong *repulsive* energy between, e.g., erythrocytes at $d = d_0$, changes to *attractive*, in contrast with the fact that Table 8.1 clearly indicates that in normal, clustered water $\Delta G_{iwi}^{TOT}$ remains significantly *repulsive* up to slightly beyond 70 Å and at d > 80 Å decreases to negligible values. The reason why such a rather short-range attraction between, e.g., human red cells (caused by a hypothetical lack of cluster formation among the molecules of liquid water) would be detrimental to the suspension stability of red cells, would be as follows:

The roughly 100 Å long glycocalyx strands that protrude from the cell membranes of human erythrocytes, are endowed with a distal end of about 10 or 20 Å long, which is exceedingly hydrophilic, whereas the major, underlying part of the strands is much less hydrophilic or even close to hydrophobic (van Oss, 1994, p. 324; 2006, p. 336). The distal, hydrophilic ends of the glycocalyx strands are what makes the exterior of blood cells such as erythrocytes as well as leukocytes themselves very hydrophilic (as determined by contact angle measurements), see, e.g., the positive value of $\Delta G_{iwi}^{TOT}$, at $d = d_0$ (as shown in Table 8.1), which is almost entirely due to the high positive value of $\Delta G_{iwi}^{AB}$. A long range (i.e., at d slightly beyond 70 Å) repulsive action which normally occurs between blood cells *in vivo*, thanks to the extended value of the decay length ($\lambda$) of liquid water of about 6 Å, assures their entirely adequate mutual repulsion and in so doing, maintains their suspension stability in blood. However, if water clusters did not exist and $\lambda$ would therefore only amount to about 2 Å, the total extent of the repulsive energy between the outer edges of two red cells (which have a diameter of about 70,000 Å) would only reach to about 19 Å. Given the normal variations in the radii of curvature of circulating blood cells, which are not perfect spheres, such cells then would often be capable of approaching one another to within slightly more than a 20 Å overlap of their glycocalices, which could give rise to an *attractive encounter* between opposing glycocalyx strands via a hydrophobic attraction between the less distal, more hydrophobic parts of their strands, thus leading to hemagglutination, i.e., to instability through clumping. (It should also be recalled that hydrophobic binding can readily occur between one hydrophobic and one hydrophilic entity; see Chapter 6, Sub-section 2.1.)

In other words, it is owing to the cluster formation in liquid water that our peripheral blood cells can remain in stable circulation.

CHAPTER TEN

# Hydration Energies of Atoms and Small Molecules in Relation to Clathrate Formation

## Contents



## Preamble

As seen in Chapters 5 and 8, when hydrophobic molecules or particles are immersed in water, they precipitate or flocculate under the strong influence of the hydrogen-bonding free energy of cohesion among the water molecules that surround them. There is, however, a curious exception to this rule: Hydrophobic atoms and small molecules tend not to combine together and precipitate when immersed in water, but instead, each single one of them surrounds itself individually with a "*cage*" of water molecules of hydration. This phenomenon is called "clathrate formation." A possible explanation for this phenomenon is as follows:

When hydrophobic particles or macromolecules (i) are immersed in water (w) their surfaces become hydrated with a free energy of hydration, $\Delta G_{iw} = \gamma_{iw} - \gamma_i - \gamma_w$, which usually is of the order of $-40$ to about $-113$ mJ/m$^2$. However when similarly hydrophobic *atoms* or *small molecules* are immersed in water,


*Interface Science and Technology*, Volume 16
ISSN 1573-4285, doi: 10.1016/S1573-4285(08)00210-X

they may undergo an *additional* free energy of hydration which is contributed by the *free energy of cohesion of water*, i.e., $\Delta G_{ww} = -2\gamma_w = -145.6$ mJ/m$^2$ (at 20 °C), so that the total free energy of hydration of such small entities, i.e., $\Delta G_{(i)}^{(hydration)} = \Delta G_{iw} + \Delta G_{ww} = \Delta G_{iw} - 146.6$ mJ/m$^2$. This permits such atoms and small molecules to avoid agglomeration driven by the very same free energy of cohesion of water which precipitates larger hydrophobic molecules and flocculates hydrophobic particles (i.e., by the attraction of the "hydrophobic effect"; see Chapter 5, Section 2, above), but instead allows each one of these atoms or small molecules to become *individually* enveloped by a "cage" consisting of cohesively hydrogen-bonded water molecules.

This process of *encaging* individual atoms or small molecules by encircling them with cohesively bound water molecules is called *clathrate formation*, where "clathrate" is derived from the Greek "kleio" (κλείω), i.e., "to close." On the other hand, larger apolar or partly polar hydrophobic molecules, macromolecules or particles are instead *collectively* surrounded and confined, driven by the free energy of cohesion of water ($\Delta G_{ww}$), to form precipitates, agglomerates or flocculates.

## 1. Free Energy of Hydration of Atoms and Small Molecules Immersed in Water

### 1.1 The $\Delta G_{iw}$ part

The free energy of hydration ($\Delta G_{iw}$) of molecules, macromolecules, particles, cells and other condensed-phase bodies (i), when immersed in water (w) is, according to the original Dupré equation (see Chapter 2):

$$\Delta G_{iw} = \gamma_{iw} - \gamma_i - \gamma_w, \tag{2.8}$$

where $\gamma_{iw}$ (see Chapter 4) is:

$$\gamma_{iw} = (\sqrt{\gamma_i^{LW}} - \sqrt{\gamma_w^{LW}})^2 + 2[\sqrt{(\gamma_i^- \gamma_i^+)} + \sqrt{(\gamma_w^- \gamma_w^+)} - \sqrt{(\gamma_i^- \gamma_w^+)} - \sqrt{(\gamma_i^+ \gamma_w^-)}]. \tag{4.1}$$

Combining Eqs. (2.8) and (4.1), one arrives at:

$$\Delta G_{iw} = -2[(\gamma_i^{LW} \gamma_w^{LW}) + \sqrt{(\gamma_i^+ \gamma_w^-)} + \sqrt{(\gamma_i^- \gamma_w^+)}]. \tag{10.1}$$

In the case where an atom or a small molecule (i) is completely apolar, $\Delta G_{iw}$ only equals $-2\sqrt{(\gamma_i^{LW}\gamma_w^{LW})}$. If molecule (i) is, for instance, cyclopentane (whose $\gamma_i^{LW}$ equals $-22.6$ mJ/m$^2$, whilst the $\gamma_w^{LW}$ value for water is 21.8 mJ/m$^2$, at 20 °C), then $\Delta G_{iw}^{LW}$ only equals $-2\sqrt{(22.6 \times 21.8)} = -44.4$ mJ/m$^2$. However, for polar molecules (i), one has to use all the right-hand terms of Eq. (10.1) to obtain $\Delta G_{iw}$, so that with even partly polar molecules the (absolute) value for $\Delta G_{iw}$ tends to be considerably higher than for completely non-polar molecules.

Meanwhile, very small, approximately spherical entities such as single atoms or small molecules, when immersed in water, can undergo an additional, second mode of attraction driven by the water molecules with which they are surrounded, i.e., the free energy of *cohesion* between the water molecules that directly encircle such atoms or small molecules. The energy of interaction caused by the free energy of attraction between these encircling water molecules may at first sight appear to be just tangentially directed around such immersed atoms or small molecules. However, in practice the energy of cohesion between the encircling water molecules rather resembles a string that is pulled tight around a small sphere, thus exerting a radial force perpendicular to the encircled atom or small molecule, oriented toward its center. The maximum size of the small molecules that can undergo this type of enhancement of their energy of hydration by the free energy of cohesion of water appears limited to about 20 Å; (Davidson, 1973; see also Section 4, below).

## 1.2 The $\Delta G_{ww}$ part

When immersed in water, the free energy of cohesion between the water molecules surrounding entity (i) is expressed as:

$$\Delta G_{ww} = -2\gamma_w = -2\gamma_w^{LW} - 2\gamma_w^{AB} = -2\gamma_w^{LW} - 4\sqrt{(\gamma_w^{+}\gamma_w^{-})} \qquad (10.2)$$

(see Chapter 2, Eq. (2.10)). At 20 °C, the Lifshitz–van der Waals (LW) part of $\Delta G_{ww}$ equals $-2 \times 21.8 = -43.6$ mJ/m$^2$ and the polar, Lewis acid–base (AB) part of $\Delta G_{ww}$ equals $-4\sqrt{(\gamma_w^{+}\gamma_w^{-})} = -102.0$ mJ/m$^2$, so that in total, $\Delta G_{ww}$ equals $-43.6 - 102.0 = -145.6$ mJ/m$^2$. Thus, $\Delta G_{ww} = -145.6$ mJ/m$^2$ can be a *constant* additional contribution to the hydration energy of atoms or small molecules when these are immersed in water, over and above the general free energy of hydration expressed as $\Delta G_{iw}$; see Eq. (10.1).

To explain the origin and the mechanism of this additional, second component of the free energy of hydration, valid only for atoms or small molecules, when these are immersed in water:

$\Delta G_{iw}$ is the free energy of direct *adhesion* between the contiguous water molecules (w) and the surface of immersed material (i) (which encloses atoms or small molecules), acting in a manner of speaking as though the water molecules are glued to the surface of the immersed entities with an energy equal to $\Delta G_{iw}$.

$\Delta G_{ww}$, on the other hand, is the free energy of *cohesion* between the water molecules of the aqueous medium which surround the atoms or small molecules and which behaves more like a string, pulled tight around each immersed atom or molecule and in so doing furnishes a large additional attractive energy of adhesion between the adjacent water molecules and the immersed atoms or small molecules. Furthermore, the large contribution by $\Delta G_{ww}$ to the free energy of hydration of small molecules and even noble gas molecules when immersed in water, can lead to the formation of *clathrate hydrates* of these small entities, see Section 4, below.

## 1.3 $\Delta G_{iw}$ and $\Delta G_{ww}$ combined

Combining Eq. (2.8, above) with Eq. (10.2), one obtains:

$$\begin{aligned}\Delta G^{(\text{Hydration})} &= \Delta G_{iw} + \Delta G_{ww} = \gamma_{iw} - \gamma_i - \gamma_w - 2\gamma_w \\ &= \gamma_{iw} - \gamma_i - 3\gamma_w \qquad (10.3)\end{aligned}$$

which, including all LW surface tension components and AB surface tension parameters (see Eqs. (10.1) and (10.2)) becomes:

$$\Delta G^{(\text{Hydration})} = -2\sqrt{(\gamma_i^{LW}\gamma_w^{LW})} - \boldsymbol{2\gamma_w^{LW}} - 2\sqrt{(\gamma_i^+\gamma_w^-)} - 2\sqrt{(\gamma_i^-\gamma_w^+)} \\ \boldsymbol{- 4\sqrt{(\gamma_w^+\gamma_w^-)}}, \qquad (10.4)$$

where the two terms printed in *bold characters* pertain to the $\Delta G_{ww}$ contribution. It should be stressed that these two terms indicated in bold characters are constant and always active, whether atoms or small molecules (i) are polar or non-polar. As indicated in Sub-section 1.2, above, these two $\Delta G_{ww}$ terms equal $-43.6 - 102.0 = -145.6$ mJ/m$^2$, at 20 °C, as they represent constant and permanent properties of water.

## 2. Hydration of Small Apolar Molecules

With completely apolar (hydrophobic) atoms or small molecules, both LW terms are present, *plus the polar term for the free energy of cohesion of water* [$=-4\sqrt{(\gamma_w^+\gamma_w^-)}$ ], *which is always active*:

$$\Delta G^{(\text{Hydration})}\ (\text{apolar}) = -2\sqrt{(\gamma_i^{LW}\gamma_w^{LW})} - 2\gamma_w^{LW} - 4\sqrt{(\gamma_w^+\gamma_w^-)}, \qquad (10.5)$$

where the two right-hand terms add up to $-145.6$ mJ/m$^2$, so that:

$$\Delta G^{(\text{Hydration})}\ (\text{apolar}) = -2\sqrt{(\gamma_i^{LW}\gamma_w^{LW})} - 145.6\ (\text{mJ/m}^2) \qquad (10.5\text{A})$$

and where $-2\sqrt{(\gamma_i^{LW}\gamma_w^{LW})}$ usually varies only between $-40$ and $-60$ mJ/m$^2$ (for hydrophobic surfaces).

For cyclopentane (a hydrophobic molecule), $-2\sqrt{(\gamma_i^{LW}\gamma_w^{LW})} = -44.4$ mJ/m$^2$ (see Sub-section 1.1, above), so that its total free energy of hydration $\Delta G^{(\text{Hydration})}$ (apolar) $= -190.0$ mJ/m$^2$ (using Eq. (10.5(A)). It is of interest to compare this value of $-190.0$ mJ/m$^2$ obtained for the total free energy of hydration of a cyclopentane molecule ($\Delta G_{iw} + \Delta G_{ww}$), with the free energy of interaction ($\Delta G_{iwi}$) between two cyclopentane molecules, when immersed in cold water, which is expressed as:

$$\Delta G_{iwi} = -2(\sqrt{\gamma_i^{LW}} - \sqrt{\gamma_w^{LW}})^2 - 4\sqrt{(\gamma_w^+\gamma_w^-)} \qquad (10.6)$$

[cf. Eq. (5.1), here containing only those terms that are applicable to completely apolar (hydrophobic) molecules or particles]. For cyclopentane (which has a $\gamma_i^{LW} = 22.6$ mJ/m$^2$ at 20 °C; see Jasper, 1972), using Eq. (10.6), $\Delta G_{iwi}$ then is equal to $-0.014 - 102.0 = -102.014$ mJ/m$^2 \approx -102.0$ mJ/m$^2$.

Thus, for *one* completely apolar molecule such as cyclopentane, the total free energy of hydration: $\Delta G^{(\text{Hydration})}$ (apolar) $= \Delta G_{iw} + \Delta G_{ww} = -190.0$ mJ/m$^2$, in water, is 1.86 times greater than the free energy of interaction ($\Delta G_{iwi} = -102.0$ mJ/m$^2$) between *two* such molecules, when immersed in water; see Table 10.1A. This mechanism, which applies only to very small entities, is of importance in considering the energetics of the formation of clathrate hydrates with single apolar molecules as guest molecules; see Section 4, below.

**Table 10.1** Comparison between the normal hydration energy of (i) ($\Delta G_{iw}$) and its extended hydration energy ($\Delta G_{iw} + \Delta G_{ww}$), with the free energy of hydrophobic attraction ($\Delta G_{iwi}$) between two molecules of (i), when immersed in water (w)

(A) **Cyclopentane (a non-polar molecule)**
$[\gamma_i^{LW} = 22.6\ mJ/m^2;\ \gamma_i^+ = \gamma_i^- = 0]$

$$\Delta G_{iw} = -44.4\ mJ/m^2$$
$$\Delta G_{ww} = -145.6\ mJ/m^2$$
$$\Delta G_{iwi} = -102.0\ mJ/m^2$$

$|\mathbf{\Delta G_{iwi}}| > |\mathbf{\Delta G_{iw}}|$
**102.0** **44.4**
**vs**
$|\mathbf{\Delta G_{iwi}}| < (|\mathbf{\Delta G_{iw}}| + |\mathbf{\Delta G_{ww}}|)$
**102.0** **(44.4 + 145.6)**

(B) **Chloroform (a partly polar molecule)**
$[\gamma_i^{LW} = 27.15\ mJ/m^2;\ \gamma_i^+ = 1.50\ mJ/m^2;\ \gamma_i^- = 0]$

$$\Delta G_{iw} = 61.0\ mJ/m^2$$
$$\Delta G_{ww} = -145.6\ mJ/m^2$$
$$\Delta G_{iwi} = -127.3\ mJ/m^2$$

$|\mathbf{\Delta G_{iwi}}| > |\mathbf{\Delta G_{iw}}|$
**127.3** **61.0**
**vs**
$|\mathbf{\Delta G_{iwi}}| < (|\mathbf{\Delta G_{iw}}| + |\mathbf{\Delta G_{ww}}|)$
**127.3** **(61.0 + 145.6)**

## 3. Hydration of Small Partly Polar Molecules

The complete free energy of hydration of very small entities is given in Eq. (10.4), above. Taking chloroform as an example, its total free energy of hydration can be derived using that equation, on the basis of the following surface-thermodynamic properties of chloroform: its $\gamma_i^{LW} = 27.15\ mJ/m^2$ and its $\gamma_i^+ = 1.50\ mJ/m^2$, whilst its $\gamma_i^- = 0$ (van Oss, 2006, p. 213), so that using Eq. (10.4) one obtains $\Delta G^{(Hydration)} = -206.6\ mJ/m^2$.

Comparing this value with the interfacial free energy of interaction between two chloroform molecules, when immersed in water ($\Delta G_{iwi}^{IF}$) and using Eq. (5.1), one finds that $\Delta G_{iwi}^{IF} = -127.3\ mJ/m^2$. Here the absolute value of $\Delta G^{(Hydration)}$ is 1.62 times greater than that of $\Delta G_{iwi}^{IF}$; see Table 10.1(B). Thus, as in the case of completely hydrophobic (apolar) molecules, here also the free energy of hydration of a partly polar molecule is significantly greater than the free energy of attraction between two such partly polar molecules, when immersed in water. This is of importance in understanding the surface-thermodynamic aspect of the mechanism of clathrate hydrate formation around small, single, partly polar molecules; see Section 4, below.

In general, partly polar molecules tend to have a higher free energy of hydration than completely apolar molecules, because with polar molecules more (attractive) terms of Eq. (10.4) are relevant than is the case with completely apolar molecules (see Eq. (10.5)). Thus, partly polar atoms or small molecules should also be able to undergo the same fate as completely apolar atoms or small molecules.

## 4. Clathrate Formation as a Hydration Phenomenon Occurring with Atoms or Small Molecules

### 4.1 The free energy of cohesion between the water molecules in liquid water only contributes to the hydration energy of immersed atoms or small molecules

It should be understood that the free energy of hydration of, for instance, a macroscopic-size flat plate, immersed in water ($\Delta G_{iw}$), is exclusively and completely described by Eq. (10.2) and the free energy of cohesion of the water adjoining such a flat plate does *not* contribute to its free energy of hydration. It is only with very small spherical, or close-to-spherical atoms or small molecules, when immersed in (cold) water, that the free energy of cohesion of the contiguous layer of water molecules surrounding such small spherical bodies, although placed tangential to the surfaces of such little spheres, will exert a radial force perpendicular to the spheres' surfaces, much as a string tied tightly around a rubber ball.

The reason for the smallness of the atoms or small molecules to which this phenomenon is limited is that even with the water molecules in liquid water, cohesive hydrogen-bond-driven attachments among the water molecules are continuously formed, and quickly broken, and formed again, all on a picosecond time scale (see Chapter 9, above; see also Luzar and Chandler, 1996). At 20 °C the average size of water clusters in liquid water is about 4.5 $H_2O$ molecules (van Oss, Giese and Good, 2002), but at colder temperatures that number is higher. It is therefore reasonable to conclude that at temperatures close to the freezing point of water, hydrogen-bonded cages of water with a cage comprising up to about 17 water molecules can form around single atoms or single small molecules (see Davidson, 1973). Beyond that size $H_2O$ molecules which are hydrogen-bonded together in liquid water become too prone to frequent dissociation to be able to maintain the integrity of a close-to-spherical cage encircling larger molecules than, e.g., cyclohexane (see Davidson, 1973; see also Table 10.1(A)).

### 4.2 Influence of $\Delta G_{iw}^{IF}$ on larger molecules or particles, immersed in water

On the other hand, systems of continuously associating and quickly dissociating polymers (e.g., clusters) of water molecules, driven by the free energy of cohesion of water, largely described by the third term of Eq. (5.1) ($\Delta G_{iwi}^{IF}$, representing the interfacial free energy of attraction between two identical molecules or particles, when immersed in water), which term is also a permanent feature of $\Delta G_{ww}$ (see Eq. (10.2)), are entirely capable of precipitating, flocculating or agglomerating vast numbers of hydrophobic molecules or particles, when immersed in water. However, the persistence of repeated brief interruptions and reassociations of such larger structures of hydrogen-bonded water molecules does not favor the formation or the integrity of clathrate hydrates encompassing multi-nanometer sized molecules or particles (see Table 10.1(A and B)).

## 4.3 Conclusion

The mechanism by which only atoms or (usually hydrophobic) small molecules, when immersed in (cold) water, can become individually encaged within a sphere of hydrogen-bonded water molecules, thus forming clathrate hydrates, is due to the fact that such clathrate configurations comprising only one guest atom or small molecule are surface thermodynamically favored; see Table 10.1(A and B). This is because in these special cases of single small guest atoms or small molecules, the free energy of cohesion between the (cold) water molecules encircling the guest atom or molecule, superimposes itself on, and in so doing amplifies, the free energy of hydration of each separate atom or small molecule, which then becomes firmly encaged as an individual clathrate hydrate.

With larger hydrophobic, or partly hydrophobic molecules or particles $\Delta G_{ww}$ only contributes (as part of $\Delta G_{iwi}$), mainly thanks to their common polar free energy of cohesion term $[-4\sqrt{(\gamma_w^+\gamma_w^-)}]$, to the attraction driving the hydrophobic effect, thus causing the precipitation, flocculation or agglomeration of larger hydrophobic molecules or particles, when immersed in water.

## 4.4 Alternative and simplified explanation

An alternative explanation of the stability of clathrates may also be considered, to the effect that once a hydrophobic atom or small molecule is surrounded by a water-cage, or clathrate, the exterior of such an aqueous clathrate (c) may be treated as a sphere of which the distal, aqueous part, interfaces with the surrounding bulk water (w). Therefore, $c = w$, its interfacial tension with the bulk water ($\gamma_{cw}$) is equal to zero, which causes the interfacial free energy of interaction between two such clathrate spheres, when immersed in water, $\Delta G_{cwc}$, also to be equal to zero. This lends complete stability (or aqueous solubility) to all aqueous clathrates; see also Chapter 7, Table 7.1).

CHAPTER ELEVEN

# THE WATER–AIR INTERFACE

## Contents



## 1. HYPERHYDROPHOBICITY OF THE WATER–AIR INTERFACE

### 1.1 Causes of the hyperhydrophobicity of the water–air interface

#### 1.1.1 $\Delta G_{air-water-air}^{IF}$

Hydrophobicity and hydrophilicity have been defined in Sub-section 1.3.1. of Chapter 5, as $\Delta G_{iw}^{IF}$, i.e., via the sign and value of the *interfacial* (IF) free energy of interaction between two molecules or particles or surfaces (i), when immersed in water (w) (where $\Delta G_{iwi}^{IF} = \Delta G_{iwi}^{LW} + \Delta G_{iwi}^{AB}$). When $\Delta G_{iwi}^{IF}$ is positive, is hydrophilic and when $\Delta G_{iwi}^{IF}$ is negative, i is hydrophobic and the degree of hydrophilicity or hydrophobicity is indicated quantitatively by the *amount* by which $\Delta G_{iwi}^{IF}$ is positive (in the case of hydrophilicity) or negative (in the case of hydrophobicity). $\Delta G_{iwi}^{IF}$ is completely defined by Eq. (5.1) and is repeated here:

$$\Delta G_{iwi}^{IF} = -2\gamma_{iw} = -2(\underbrace{\sqrt{\gamma_i^{LW}} - \sqrt{\gamma_w^{LW}}}_{1})^2 - 4[\underbrace{\sqrt{(\gamma_i^+\gamma_i^-)}}_{2} + \underbrace{\sqrt{(\gamma_w^+\gamma_w^-)}}_{3} - \underbrace{\sqrt{(\gamma_i^+\gamma_w^-)}}_{4} - \underbrace{\sqrt{(\gamma_i^-\gamma_w^+)}}_{5}]. \quad (11.1)$$


*Interface Science and Technology*, Volume 16
ISSN 1573-4285, doi: 10.1016/S1573-4285(08)00211-1

When material (i) is *air* (henceforth indicated by a subscript, "a"), terms 2, 4 and 5 are zero, so that:

$$\Delta G_{awa}^{IF} = -2\gamma_w^{LW} - 4\sqrt{(\gamma_w^+\gamma_w^-)}. \qquad (11.2)$$

Thus, according to Eq. (11.2), the interfacial free energy of interaction between two air bubbles (a), separated by water (w), $\Delta G_{awa}^{IF}$, *is equal to the LW plus the AB free energy of cohesion of water*, or $-43.6 - 102.0 = -145.6$ mJ/m$^2$ which indicates a strong attraction, signifying that air, at the water–air interface, is exceedingly hydrophobic, so that two air bubbles, immersed in and separated by water, will attract one another quite strongly. For all *condensed-phase materials* (air, at room temperature and ambient pressure is *not* a condensed-phase material) term 1 of Eq. (11.1) is necessarily smaller than 43.6 mJ/m$^2$ and can be as small as zero (e.g., for octane while immersed in water, because octane has virtually the same $\gamma^{LW}$ as water) so that for *air bubbles*, immersed in water, $\Delta G_{awa}^{LW}$ can be up to 43.6 mJ/m$^2$ stronger than for a condensed-phase material (such as octane), when immersed in water. In addition to $\Delta G_{awa}^{LW}$, there is the polar free energy of cohesion of water: $\Delta G_{awa}^{AB} = -102$ mJ/m$^2$, which is always present, being a permanent property of water (see Chapter 5, Sub-section 2.1, above).

The hydrophobicity of the water–air interface is therefore greater than that of any condensed-phase material (van Oss, Docoslis and Giese, 2002; van Oss, Giese and Docoslis, 2005).

### 1.1.2 The contact angle measured with water on air

Reiterating the complete apolar *plus* polar Young–Dupré equation from Chapter 2 (Eq. (2.15B)), which expresses the relation between a contact angle measured with a drop of a given liquid on flat solid material, it is possible to observe the value of the contact angle ($\theta$) one can obtain with, e.g., water (w), on a flat layer of air (a):

$$(1 + \cos\theta)\gamma_w = 2[\sqrt{(\gamma_a^{LW}\gamma_w^{LW})} + \sqrt{(\gamma_a^+\gamma_w^-)} + \sqrt{(\gamma_a^-\gamma_w^+)}\,]. \qquad (11.3)$$

Now, because $\gamma_a^{LW}$, $\gamma_a^+$ and $\gamma_a^-$ are all zero, Eq. (11.3) is reduced to:

$$(1 + \cos\theta)\gamma_w = 0 \qquad (11.4)$$

so that $\cos\theta = -1$ in the case of contact angles measured with drops of water on a surface of air, theoretically yielding: $\theta = 180°$; see Figure 11.1. Onda et al. (1996) proved that it is experimentally possible to approach $\theta = 180°$ quite closely with a drop of water, deposited on a flat, ultra-porous surface consisting of fractally repeating air holes in a foam consisting of an alkyl ketene dimer, which presented 99.2% of its surface as air (see van Oss, Giese and Docoslis, 2005). Although it has not been possible to date to obtain a water contact angle of actually 180° on a layer of air, Onda et al. (1996) achieved a contact angle of 174° on their fractal foam, of which a photograph appeared on the cover of the 1996 Langmuir issue which contained Onda et al.'s publication.

One may also write Eq. (11.3) (see Eq. (2.15A)) as:

$$(1 + \cos\theta)\gamma_w = -\Delta G_{aw}, \qquad (11.5)$$

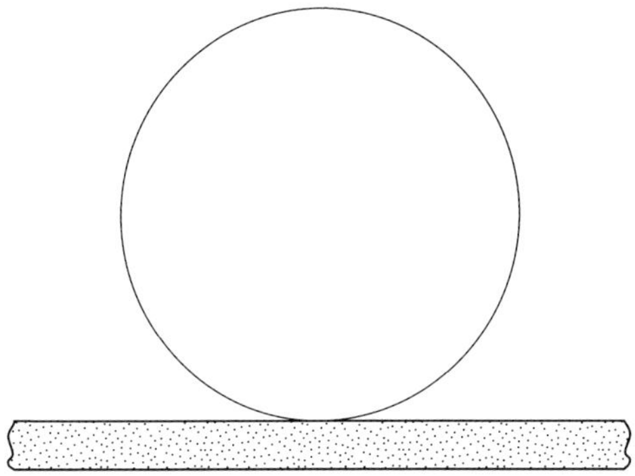

**Figure 11.1** Schematic presentation (approximately 10× enlarged) of an exceedingly porous, hydrophobic, solid foam plaque upon which a drop of water is placed, making a contact angle of (ideally) 180°, which in actual practice is hard to reach. However, Onda et al. (1996) achieved a water contact angle of 174° on a fractal foam plaque of hydrophobic material, with 99.2% surface porosity (see text).

where $\Delta G_{aw}$ is the free energy of hydration of the water–air interface. From Eqs. (11.4) and (11.5), it follows that $\Delta G_{aw} = 0$, signifying that the water–air interface (at the air side) is not hydrated at all. From Eqs. (11.4) and (11.5) it can be seen that the air–water interface is the only interface (together with other water–gas interfaces and the water–vacuum interface) that is genuinely *hydrophobic* (=water-fearing). In contrast with this, low-energy condensed-phase materials, such as Teflon ($\gamma_{Teflon} = 17.9$ mJ/m$^2$, van Oss, 1994, 2006) are by most people erroneously called "hydrophobic." Teflon (t) has a free energy of hydration ($\Delta G_{tw}$), which in the case of Teflon is equal to: $\gamma_{tw} - \gamma_t - \gamma_w = 51.2 - 17.9 - 72.8 = -39.5$ mJ/m$^2$ (see Eq. (2.8)) which means that Teflon binds water fairly strongly and thus is far from "hydrophobic", *sensu stricto*; see also Hildebrand (1979).

## 1.2 Hyperhydrophobicity of the water–air interface as the sole cause of the increased value of the water contact angle when measured on rough solid surfaces

The increase in contact angles found with drops of water on rough solid surfaces is solely a consequence of the presence of air and the proportion of the interstitial surface area of air between protruding roughness peaks, as compared to the total surface area of the rough surface. As shown in the preceding Sub-section 1.1.2, above, in the case where the % air surface in contact with the bottom of the contact angle drop tends toward 100%, the contact angle measured with water tends toward 180°. The increased water contact angles found on rough surfaces with many concave water interstices was studied by Cassie and Baxter (1944) and by Cassie (1948); see also van Oss, Giese and Docoslis (2005). Cassie and Baxter were mainly interested in the impermeability for water drops of woven textiles destined for the manufacture of water-proof rain coats (in French aptly called "imperméables"). The need to calculate the influence of the proportion of exposed air surface area on rough surfaces prompted Cassie (1948) to utilize what is now generally called the Cassie equation:

$$\cos\theta_A = f_1 \cos\theta_1 + f_2 \cos\theta_2, \qquad (11.6)$$

where subscript A indicates that $\theta_A$ is the contact angle obtained experimentally with a drop of water on the total "aggregate" rough surface and where $f_1$ is the fraction of the surface represented by the solid support material and $f_2$ is the fraction of the surface represented by air, such that $f_2 = 1 - f_1$. Furthermore, $\theta_1$ is the contact angle obtainable on a surface consisting of a sample of smooth, uninterrupted support material and $\theta_2$ is the contact angle which is theoretically obtainable on the water–air interface (i.e., $\theta_2 = 180°$ and $\cos\theta_2 = -1$). With the help of Cassie's equation it was possible to determine that the proportion of air surface on Onda et al.'s (1996) porous fractal surface was 99.2% (van Oss, Giese and Docoslis, 2005).

### 1.3 Hyperhydrophobicity of the water–air interface as the basis of flotation as a separation method

The introduction of air bubbles near the bottom of a vessel filled with water causes the air bubbles to rise to the top of the water column and in so doing presents a rather large hydrophobic surface area formed by the total number of rising air bubbles at any given point in time. The hydrophobicity of the rising air bubbles at the water–air interface can be used to attach to various mineral particles when suspended in water and thus to separate them selectively on their way up. This separation method is called "flotation." The one drawback of having a great many bubbles rising at the same time is that they strongly attract one another in water, which causes them to fuse together, thus drastically decreasing their total surface area. The fusing of bubbles can be diminished or even prevented by adding some surfactant to the water. This is, however, accompanied by the less desirable property of the added surfactants in that they cause a considerable decrease in the hydrophobicity of the rising bubbles' water interface, brought about by migration of the hydrophobic tails of the added surfactant molecules toward the air-side of the bubbles' water–air interfaces, whilst the hydrophilic heads of the surfactant molecules protrude into the water side of that interface, which seriously diminishes the bubbles' hydrophobicity as approached from the water side. The optimal efficacy of flotation with the help of rising air bubbles as a purification method must therefore be based on a compromise between a maximum total bubble surface area brought about by the addition of a surfactant *versus* a minimum decrease in the hydrophobicity of the bubbles' total water–air interface area accessible from the water-side, caused by the admixture of surfactant.

## 2. The $\zeta$-Potential of Air or Gas Bubbles in Water

There are only a few reports of measured electrophoretic mobilities and thus of $\zeta$-potentials of air (or $N_2$) bubbles in pure water, without the presence of surfactants or proteins, which prevent the bubbles from coalescing but in so doing also makes it impossible to determine the bubbles' *own* $\zeta$-potential at the water–air interface. However, Graciaa et al. (1995) managed to measure the $\zeta$-potential of air bubbles in deionized water in a rotating tube (aimed at preventing the bubbles to rise vertically by rotating the tube around its horizontal axis). They found a $\zeta$-potential of about $-65$ mV. On the other hand, Craig et al. (1993), using bubbles filled with $N_2$, found rather surprisingly that the addition of *salt* to the water could serve to *prevent* the attraction (and hence the fusion) among bubbles, although in all other known aqueous particle suspensions the addition of salts *favors* interparticle attraction (cf. van Oss et al., 2005; see also Chapter 3, Sub-section 6.4). It should however be recalled that without the presence of at least some tiny amounts of, e.g., NaCl, in pure water, the measurement of reproducible electrophoretic mobilities of entities which are immersed in that water is not easy to do.

In general, the experimental complexities involved in preventing the fusion of rising air bubbles in water makes it difficult to determine their $\zeta$-potential without the stabilizing presence of surfactants and *with* the addition of surfactants one can only find the $\zeta$-potentials of the added surfactants and not the $\zeta$-potential of the water–air interface.

Meanwhile one should, for the time being, accept the $\zeta$-potential of $-65$ mV for the water–air interface found by Graciaa et al. (1995) and maybe further investigate the puzzling influence of added salt on the $\zeta$-potential of immersed air bubbles (cf. Craig et al., 1993).

# 3. Repulsion versus Attraction of Various Solutes by the Water–Air Interface

## 3.1 Determination of repulsion or attraction of solutes by the water–air interface

*There are only two categories of solutes which dissolve in water:*

*Category 1* solutes are *repelled* by the water–air interface; they are usually hydrophilic or close-to-hydrophilic.

*Category 2* solutes are *attracted* to the water–air interface *via* their hydrophobic moiety. (Such solutes are usually amphiphilic, i.e., they have a hydrophobic and a hydrophilic moiety.)

The only known solute that would be very little influenced by the water–air interface is probably $D_2O$, whose relevant physico-chemical and surface-thermodynamic properties are so close to those of $H_2O$ that as a solute, $D_2O$ would be neither repelled nor attracted to the water–air interface to any significant degree.

The *degree of either repulsion or attraction of different solutes* by the water–air interface can be *calculated* using Eq. (6.1) for $\Delta G_{swa}$, where (s) is the immersed solute and (a) is air (for which $\gamma_a^{LW} = \gamma_a^+ = \gamma_a^- = 0$) and where the surface-thermodynamic properties of the solute (s) (i.e., $\gamma_s^{LW}$, $\gamma_s^+$ and $\gamma_s^-$) have to be measured (see Chapter 2, Sub-section 1.6). In addition, the known surface properties of water (w) have to be taken into account (cf. Chapter 2, Table 2.1).

On the other hand, the *attraction of hydrophobic solutes* to the water–air interface can be *directly demonstrated* by the decrease in the apparent surface tension of aqueous solutions as a function of increasing solute concentrations while noting the occurrence of a plateau value once a given solute concentration is reached (Absolom et al., 1981a, 1981b), which is the point of maximum aqueous solubility of the solute, at which point upon further addition of solute, the excess solute henceforth finds a refuge in the formation of micelles, i.e., the "solubility" of, e.g., surfactant molecules is equivalent to the critical micelle concentration (cmc), see Chapter 7, Sub-section 2.3; see also this chapter, Sub-section 3.3.1, below.

## 3.2 *Repulsion* of hydrophilic or near-hydrophilic solutes by the water–air interface

In contrast with the *attraction of hydrophobic solutes* to the water–air interface, which is very noticeable (see Sub-section 3.3, below), the *repulsion of hydrophilic or close-to-hydrophilic solutes* by the water–air interface cannot be observed directly, but is easily determined via the $\gamma_s^{LW}$, $\gamma_s^+$ and $\gamma_s^-$ values of the solute in question, and calculation of the resulting $\Delta G_{swa}$ value, using Eq. (6.1, Chapter 6), upon which it will be seen that with such solutes a positive $\Delta G_{swa}$ value ensues, denoting a repulsion between the solute molecules and air, separated by water. In almost all cases $\Delta G_{swa}^{LW}$ (and often also $\Delta G_{swa}^{AB}$) will be found to be positive (i.e., repulsive) with hydrophilic or close-to-hydrophilic solutes.

With $\Delta G_{swa}^{LW}$ this is the case because air has a zero value for $\gamma_a^{LW}$, whilst for water $\gamma_w^{LW} = 21.8$ mJ/m$^2$ and for most solutes $\gamma_s^{LW}$ has a value between 25 and 45 mJ/m$^2$, so that the $\gamma_w^{LW}$ value for water always lies between that of the solute and the zero value for air, which leads to an LW repulsion between the solute and air when separated by water; see Chapter 1, Sub-section 4.1 and Chapter 6, Sub-section 1.1. This may be elucidated as follows:

$$\Delta G_{swa}^{LW} = 2[\sqrt{(\gamma_s^{LW}\gamma_w^{LW})} + \sqrt{(\boldsymbol{\gamma_a^{LW}}\gamma_w^{LW})} - \sqrt{(\gamma_s^{LW}\boldsymbol{\gamma_a^{LW}})} - \gamma_w^{LW}] \tag{11.7}$$

(cf. Eq. (6.1), the LW part). When a = 0 (see the *bold* terms in Eq. (11.7)), Eq. (11.7) is reduced to:

$$\Delta G_{swa}^{LW} = 2\sqrt{(\gamma_s^{LW}\gamma_w^{LW})} - 2\gamma_w^{LW} \tag{11.8}$$

or:

$$\Delta G_{swa}^{LW} = 2\sqrt{(\gamma_s^{LW} \times 21.8^{\star})} - 43.6^{\star} \quad (\star \text{ both in mJ/m}^2). \tag{11.8A}$$

Thus, $\Delta G_{swa}^{LW}$ is always positive (i.e., repulsive) when $\gamma_s^{LW} > \gamma_w^{LW}$ (where $\gamma_w^{LW} = 21.8$ mJ/m$^2$).

However a positive $\Delta G_{swa}^{LW}$ is not the only factor governing the outcome of repulsion *versus* attraction. The repulsion between *polar* solutes, immersed in water, and the water–air interface (in addition to the influence of $\Delta G_{swa}^{LW}$, discussed above) can be further enhanced by a positive value for $G_{swa}^{AB}$. To determine the conditions under which $G_{swa}^{AB}$ is positive, it is useful to refer to the AB part of

Eq. (6.1, see Chapter 6):

$$\Delta G_{swa}^{AB} = 2[\sqrt{\gamma_w^+}(\sqrt{\gamma_s^-} + \sqrt{\gamma_a^-} - \sqrt{\gamma_w^-}) + \sqrt{\gamma_w^-}(\sqrt{\gamma_s^+} + \sqrt{\gamma_a^+} - \sqrt{\gamma_w^+}) - \sqrt{(\gamma_s^+\gamma_a^-)} - \sqrt{(\gamma_s^-\gamma_a^+)}\,], \quad (11.9)$$

where all terms with a subscript "a" are equal to zero, so that in all cases describing interactions including the water–air interface, Eq. (11.9) is reduced to:

$$\Delta G_{swa}^{AB} = 2[\sqrt{\gamma_w^+}(\sqrt{\gamma_s^-} - \sqrt{\gamma_w^-}) + \sqrt{\gamma_w^-}(\sqrt{\gamma_s^+} - \sqrt{\gamma_w^+})]. \quad (11.9A)$$

A comparison between Eqs. (11.9) and (11.9A) shows that a positive value for $\Delta G_{swa}^{AB}$ is favored by the elimination of the last two (negative) right-hand terms of Eq. (11.9), because everything containing subscript "a" is zero. Further scrutiny of Eq. (11.9A) shows that another configuration favoring a positive value for $G_{swa}^{AB}$ occurs when:

$$\gamma_s^- > \gamma_w^-$$

as well as when:

$$\gamma_s^+ > \gamma_w^+$$

(where the latter occurs exceedingly rarely, if ever).

This is the case with the sugars, glucose and sucrose: *glucose*: $\gamma_{gl}^{LW} = 42.2$; $\gamma_{gl}^+ = 34.5$ and $\gamma_{gl}^- = 85.6$ mJ/m$^2$ and *sucrose*: $\gamma_{su}^{LW} = 41.6$, $\gamma_{su}^+$ 28.5 and $\gamma_{su}^- = 88.0$ mJ/m$^2$ (these values in both cases are only valid for these sugars when dissolved in water; see Docoslis et al., 2000, but not for dried sugars; see van Oss, 1994, 2006). In view of these data and of the inequalities shown above, both sugars have a positive $\Delta G_{swa}^{AB}$ as well as a positive $\Delta G_{swa}^{LW}$. In aqueous solutions these sugars therefore not only cause an increase in the surface tension of water, but they also are repelled away from the water–air interface. Using the extended DLVO approach (see Chapter 3) and taking the $\Delta G_{swa}^{LW}$ and $\Delta G_{swa}^{AB}$ values into account (but disregarding $\Delta G_{swa}^{EL}$ as negligible, being less than 1 kT for, e.g., glucose, at $d = d_0$), it is possible to estimate the thickness of the sugarless void between the water–air interface and the dissolved sugar molecules in the bulk liquid. An essentially sugarless void may be taken to exist between the water–air interface and the place where the repulsion energy reaches $\Delta G_{swa} = \Delta G_{swa}^{LW} + \Delta G_{swa}^{AB} \approx +10$ kT, which is a *significant repulsion*; see, e.g., Overbeek (1977). The resumption of practically complete bulk concentration may be estimated to begin at a distance from the water–air interface where the repulsion energy ($\Delta G_{swa}$) is down to about +1 kT, indicating the approximate level of the normal energy of Brownian motion.

The sliver of water which is largely depleted of carbohydrates due to a repulsion of +10 kT represents the thickness of a band of practically pure water at the water–air interface which is only of the order of between 1 to at most 2.5 Å, whilst the total thickness, d, between the water–air interface, through a gradient of increasing carbohydrate concentration, up to the point of essentially bulk carbohydrate level, is somewhat under 10 Å.

There is a difference between the behavior in aqueous solution of single sugars (e.g., glucose or sucrose) and that of their polymers. For instance, dextran, although practically a glucose polymer (it is a polymer of maltose, which is a dimer of glucose) is quite different from either glucose or sucrose. Dissolved glucose and sucrose have high $\gamma^+$ as well as very high $\gamma^-$ values, which gives them a high $\gamma^{AB}$ value which in turn causes them to have a high polar free energy of cohesion ($\Delta G_{sws}{}^{AB}$), leading to a strong continuous attraction between these molecules' freely available electron-acceptors and electron-donors. Dextran on the other hand has a $\gamma^{AB}$ of only 21.35 mJ/m$^2$, as compared to a $\gamma^{AB}$ of 108.7 mJ/m$^2$ for its monomer, glucose (due to the elimination of almost all freely exposed electron-acceptors and electron-donors of its monomers, through the process of covalent polymerization; see Docoslis et al., 2000). Thus, contrary to glucose, dextran does not increase the surface tension of water, it *decreases* it (water has a total surface tension of 72.8 mJ/m$^2$ whilst glucose has a total surface tension of 63.6 mJ/m$^2$ (see Docoslis et al., 2000). However, largely as a consequence of dextran's $\Delta G_{swa}{}^{LW}$ of +17 mJ/m$^2$, which exceeds its negative $\Delta G_{swa}{}^{AB}$ of −11.5 mJ/m$^2$, it still is repelled by the water–air interface with a $\Delta G_{swa} = +5.5$ mJ/m$^2$.

Virtually all solutes which *increase* the surface tension of water (e.g., sugars, many salts) are repelled by the water–air interface. In addition, some solutes which *decrease* the surface tension of water (e.g., dextran) are also repelled by the water–air interface. Meanwhile, it should be noted that low-molecular weight carbohydrates such as glucose and sucrose are not hydrophilic *sensu stricto*, because their $\Delta G_{sws}{}^{IF}$ value is *negative*, which is also the case with all solutes (sugars as well as salts) that have a *finite*, well-determined aqueous solubility (cf. Chapter 5, Sub-section 1.3.3 and Eq. (5.2)). Dextran (when dissolved in water) on the other hand has a *positive* $\Delta G_{sws}{}^{IF}$ and is therefore *genuinely hydrophilic* by definition and infinitely soluble in water, in practice only limited by the high viscosity of very concentrated dextran solutions.

### 3.2.1 Summary of the properties of solutes contributing to their repulsion by the water–air interface

There are two main categories of solutes that are *repelled* by the water–air interface:

***A. Solutes which in aqueous solution increase the surface tension of water and are repelled by the water–air interface*** These solutes generally have a $\gamma^{LW}$ as well as a $\gamma^{AB}$ higher than that of water, when in solution, i.e., they are strongly dipolar. They include:

Simple *sugars* such as *glucose* and *sucrose* and also some of the smaller molecular weight *amino acids* such as *glycine* (Docoslis et al., 2000); Many *salts* such as: NaCl, KCl, NaBr, $MgCl_2$, $NaNO_3$, $Na_2SO_4$, $Na_2CO_3$, $Al_2(SO_4)_3$ (see Weast, 1970/71, p. F28, for the increase in the surface tension of water caused by the admixture of these salts).

Both the above-mentioned sugars as well as the above-mentioned salts [with, in addition, $(NH_4)_2SO_4$] increase the surface tension of water and thus also increase its free energy of cohesion which in turn enhances the hydrophobizing capacity of the aqueous solution (see Chapter 5, Sub-section 2.2.1). In the cases of salts this phenomenon is also called "salting out", because when added to, e.g., an aqueous protein solution, the hydrophobizing action resulting from the addition of considerable amounts of salt [typically plurimolar amounts of $(NH_4)_2SO_4$] favors the

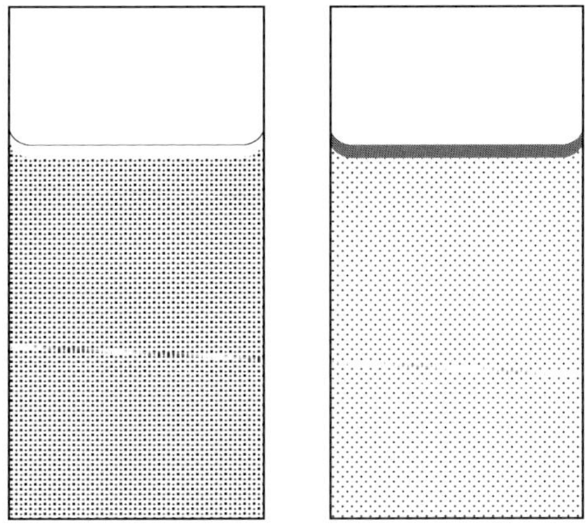


**Figure 11.2** Schematic depiction of two vessels, where vessel A shows the behavior of *Category 1* and vessel B shows the behavior of *Category 2* solutes in aqueous solution. Vessel A contains an aqueous solution of a hydrophilic or close-to-hydrophilic, *high* surface tension solute such as a sugar or a salt. Vessel B contains an aqueous solution of a *low* surface tension amphiphilic solute with a hydrophobic as well as a hydrophilic moiety, such as an ionic or a non-ionic surfactant. In vessel A, *Category 1* solute molecules are *repelled* by the water–air interface, leaving an approximately nanometer-sized, largely solute-depleted layer of just water, adjoining the water–air interface. Such high-surface tension solutes (e.g., sugars or salts) cause an *increase* in the apparent (measured) surface tension of water but due to the presence of the thin depletion zone, that measured increase in surface tension is smaller than the actual (but not easy to measure) increase in surface tension of the bulk solution. However, other solutes of *Category 1*, while still being repelled by the water–air interface, cause a *decrease* in the measured surface tension of their aqueous solution, e.g., the low-surface tension glucose polymer, dextran (see text; see also Docoslis et al., 2000). In vessel B, *Category 2* solute molecules are *attracted* to the water–air interface, typically depositing a layer of surfactant molecules at that interface of approximately the thickness of at least one surfactant molecule, with its hydrophobic tail protruding into the air-side of the water–air interface, whilst its hydrophilic moiety remains in the water. Very low concentrations of *Category 2* solute (surfactant) molecules can cause a large decrease in the apparent (measured) surface tension of the solution (Lange, 1967).

insolubilization (through hydrophobization of the dissolved protein, e.g., serum albumin), especially at the protein's isoelectric pH (see Chapter 5, Sub-section 2.2.1; see also Chapter 12, below).

In general, added salts tend to raise the apparent surface tension of water to maximally 10 to 15% and with added sucrose, only about 4.4% (Weast, 1970/71). It would thus appear that during measurement of the surface tensions of salt or sugar solutions, the retreat of the solute away from the water–air interface (see Figure 11.2A), causes a significant underestimation of the surface tension of the bulk solution.

A hypothesis by Chaudhury and Good (1985) that the *increase* in surface tension of aqueous sucrose solutions would be due to the (van der Waals) repulsion between the water–air interface and the dissolved sucrose molecules is not tenable. Given the mechanism of, e.g., liquid drop shape as a force balance (see Chapter 2, Sub-section 1.6.2), a surrounding shell of water of lower surface tension than the inner, bulk aqueous solution, which together constitute the drop, can only *decrease* the apparent measured surface tension of the drop. Chaudhury and Good (1985) proposed no mechanism in support of their conjecture.

High concentrations of the above-mentioned salts, as well as the two sugars, glucose and sucrose, also favor destabilization of aqueous particle suspensions (Docoslis et al., 2000; see also Chapter 5, Sub-section 2.2.1).

***B. Solutes which in aqueous solutions decrease the surface tension of water and are repelled by the water–air interface*** Section A (see above) comprises the most important category of solutes which are repelled by the water–air interface. However, among those (usually quite hydrophilic) solutes which *decrease* the surface tension of water there are some which are nonetheless repelled by the water–air interface. Dextran is an example of these, or more precisely, *dextran in aqueous solution*. It would appear that two requirements have to be satisfied for such a dissolved polysaccharide to be repelled by the water–air interface:

1. It needs to have a high $\gamma_s^{LW}$, which is the case with dextran, with a $\gamma_s^{LW}$ of 42 mJ/m$^2$, thus achieving a respectable repulsive $\Delta G_{swa}^{LW} = +17$ mJ/m$^2$.
2. It needs to have a high $\gamma_s^{-}$ (with dissolved dextran $\gamma_s^{-} = 57$ mJ/m$^2$), as well as a modest but non-negligible $\gamma_s^{+}$, which in the case of dissolved dextran equals $\gamma_s^{+} = 2.0$ mJ/m$^2$. These two combined suffice to yield a $\Delta G_{swa}^{AB}$ value of only $-11.5$ mJ/m$^2$ which is smaller than the $\Delta G_{swa}^{LW}$ of $+17$ mJ/m$^2$, thus leaving a residual repulsive $\Delta G_{swa}^{IF}$ $(=\Delta G_{swa}^{LW} + \Delta G_{swa}^{AB}) = 17 - 11.5 = +5.5$ mJ/m$^2$. If dissolved dextran were $\gamma_s^{-}$ monopolar, the above sum would have been equal to: $17 - 51 = -34$ mJ/m$^2$, thus causing the water–air interface to attract dextran molecules, which is not the case.

## 3.3 Solutes which comprise hydrophobic components are strongly attracted to the water–air interface

### 3.3.1 Surfactants

There is a large category of solutes that are attracted by the water–air interface. These are compounds which contain a hydrophobic moiety, in addition to a hydrophilic component. Solutes which are *attracted* to the water–air interface (see Figure 11.2B), typically *drastically decrease* the surface tension of water, upon their admixture to water in even quite small amounts. For example upon the addition of only $10^{-5}$ molar tetra-oxyethylene *n*-dodecanol to water, its aqueous solution's surface tension *decreases* from 72.8 mJ/m$^2$ (i.e., the surface tension of pure water at 20 °C) to approximately 27 mJ/m$^2$ (Lange, 1967), a 63% decrease. Upon the addition of more tetra-oxyethylene *n*-dodecanol to the initial $10^{-5}$ M solution, the measured surface tension of the aqueous solution does not decrease any further and a plateau value of about 26 to 27 mJ/m$^2$ persists indefinitely indicating that a saturation of the non-ionic surfactant molecules occupying the water–air interface had been reached, while the surplus surfactant continued to accumulate in the bulk liquid. At the point where the admixture of tetra-oxyethylene *n*-dodecanol reached a plateau value of $10^{-5}$ M, its maximum solubility as a monomolecular solute was reached at the same time and further addition of the surfactant to the bulk liquid gave rise to its insolubilization. However with surfactants this does not result in precipitation of the excess solute, but instead of classical precipitate formation, the surplus surfactant molecules organize themselves into "micelles." These are larger than monomolecular structures, comprising ≈50 to 100, or for non-ionic surfactants even thousands of monomeric units per micelle (Becher, 1967). They organize themselves in spherical or ellipsoidal or cylindrical units, in such a manner that while suspended in water, the hydrophilic chains (or "heads") of each monomeric molecule project themselves into the water phase, whereas the hydrophobic chains are internalized inside the micellar body. Thus, stable micelle *suspensions* in water may be equally legitimately considered as micelles in aqueous *solution* and the critical micelle concentration (*cmc*) may be equated with the solubility of surfactants [see, e.g., van Oss and Good (1991) and van Oss and Costanzo (1992); see also Chapter 7, Sub-sections 2.3.2–2.3.4]. The large decrease in the measured surface tension of water following the addition of surfactant molecules is only apparent: what one actually measures is the strong influence of the low surface tension of the low-energy (hydrophobic) surfactant moieties that line the water–air interface, on the total liquid surface tension measurement procedure. The preferential attraction of just the hydrophobic part of the added surfactant molecules to the water–air interface, in conjunction with the fact that said interface readily becomes completely saturated with these hydrophobic moieties at the very surface at which one measures the surface tension of the whole (now extremely anisotropic) liquid, produces the exceedingly biased appearance of a dramatic lowering of the surface tension of the whole liquid, while in reality one measures the surface tension of just a tiny unrepresentative outer sliver of the total body of liquid of which the inner bulk has altogether different properties.

### 3.3.2 Globular proteins

Globular proteins such as human serum albumin (HSA) or serum immunoglobulin-G (IgG) (see also Chapter 7, Sub-section 3.3), also are attracted to the water–air interface when in aqueous solution and in so doing, e.g., HSA causes a decrease of 17 to 18% in the apparent surface tension of the aqueous solvent (Absolom, van Oss et al., 1981b). These so-called globular proteins (so-called even though HSA is V-shaped and IgG is Y-shaped) differ from surfactants because, although they also have hydrophilic and hydrophobic components under *in vivo* conditions (i.e., their surfaces are hydrophilic while dissolved in an aqueous medium, away from any water–air interface) their hydrophobic moieties are *internalized inside each molecule* and their hydrophilic interface with water surrounds them virtually completely over their entire surface area adjoining the aqueous medium. In other words, under *in vivo* conditions globular blood serum proteins closely resemble *micelles*. However, when for example HSA molecules find themselves at a water–air interface, their tertiary structure changes to allow the interior hydrophobic peptides to reorient themselves outward, toward the hydrophobic water–air interface. This happens for instance during the process of air-drying HSA from an initial aqueous solution. The dried HSA surface then becomes hydrophobic (in the case of air-dried HSA $\Delta G_{sws} = -11.6$ mJ/m$^2$), although the original hydrated HSA (with one molecular layer of water of hydration) is extremely hydrophilic ($\Delta G_{sws}^{IF} = +85.3$ mJ/m$^2$).

[These results are based on the following surface-thermodynamic data: air-dried HSA: $\gamma^{LW} = 41.1$; $\gamma^{+} = 0.002$ and $\gamma^{-} = 20$ mJ/m$^2$, whilst for the original native HSA: $\gamma^{LW} = 26.6$, $\gamma^{+} = 0.003$ and $\gamma^{-} = 87.5$ mJ/m$^2$, all at pH 7.0 (see van Oss, 1994, p. 178; 2006, p. 219; see also Chapter 7, Sub-section 3.3 and Chapter 11, Sub-section 3.2)]. Upon reimmersion in water, HSA molecules first rehydrate and relatively slowly resume their original native hydrophilic monomolecular micelle-like structure, after which they completely redissolve in water. [Contrary to an earlier statement by Absolom, van Oss et al. (1981b), air-dried HSA is not "reversibly or irreversibly", but usually only *reversibly* denatured by the air-drying procedure, at room temperature. This reversible denaturation through air-drying HSA can be avoided by freeze-drying it instead.]

## 4. Inadvisability of Using Aqueous Solutions for the Measurement of Contact Angles

In Chapter 2, Sub-section 1.6.2, an explanation is given of the fundamental role of the shape of a drop of liquid, L, sessile on the surface of flat, smooth solid, S, in conjunction with the Young–Dupré equation. The shape of the drop, expressed as the cosine of the contact angle, is likened to a force balance in which the equilibrium between the free energy of *cohesion of the liquid* of the drop and the free energy of *adhesion between the liquid of the drop and the solid* upon which it has been deposited permits the determination of the surface properties of the solid.

This use of the contact angle a liquid makes when sessile upon a solid presupposes that the liquid is homogeneous and that its outermost layer at the liquid–air interface has exactly the same composition as the bulk liquid. If this is not the case one no longer knows the free energy of cohesion of the whole liquid making up the drop. Thus, significantly anisotropic liquids with different compositions of the liquid at the liquid-air interface and the bulk liquid, should not be used in contact angle determinations, because the real free energy of cohesion of the entire drop is no longer known. Therefore: "*Thou shalt not measure contact angles with composite liquids or solutions*!"

With *Category 2 solutes* (see Sub-section 3.1, above) this interdiction is absolute at practically all solute concentrations. However with *Category 1 solutes* this interdiction is less severe: for instance if one needs to measure contact angles on layers of mammalian cells with a physiological buffered salt solution (PBS), with an ionic strength of 0.15, the surface tension of the water in which the salt is dissolved (i.e., 72.8 mJ/m$^2$ at 20 °C) only increases to about 73.1 mJ/m$^2$, or only about 0.4%. The errors which this would cause are of the same order of magnitude as the statistical errors that usually occur in normal contact angle measurements, so that in such cases the use of fairly dilute salt solutions may be condoned.

CHAPTER TWELVE

# Influence of the pH and the Ionic Strength of Water on Contact Angles Measured with Drops of Aqueous Solutions on Electrically Charged, Amphoteric and Uncharged Surfaces

## Contents



## 1. Influence of the pH of Water on Electrically Charged and Amphoteric Surfaces

In this chapter the heading "Electrically Charged Surfaces" pertains to surfaces manifesting an electrical surface potential when immersed in water, which remains of the same sign of charge at all pH values of the aqueous medium. "Amphoteric Surfaces" on the other hand are surfaces which are positively charged at low pH

*Interface Science and Technology*, Volume 16
ISSN 1573-4285, doi: 10.1016/S1573-4285(08)00212-3

values and negatively charged at high pH values of the aqueous medium in which they are immersed.

## 1.1 Electrically charged surfaces

As mentioned above, "electrically charged surfaces" are here defined as surfaces that have the same sign of charge at all pH values of the aqueous medium with which they are in contact or in which they are immersed. Taking clean glass (i.e., a microscope slide) as an example of a (negatively) charged surface, contact angles ($\theta_w$) with drops of water of different pH values have been measured at 1.94, 6.1 and 12.8 (van Oss and Giese, 2005). At pH 1.94, $\theta_w = 9.7°$ and at pH 6.1, $\theta_w = 10.5°$; these two values were not found to be statistically different. However at pH 12.8, $\theta_w = 6.8°$, which was found to be significantly different from $\theta_w$ at pH 1.94 as well as at pH 6.1 (p-values 0.0002 and zero, respectively). This shows that (negatively charged) clean glass in contact with water is hydrophilic at pH's 1.94 and 6.1, but that it is even more hydrophilic at pH 12.8. Therefore, the contact angles measured on a (negatively charged) material like clean glass, which are low under acid and almost neutral conditions, become even lower when measured with strongly alkaline drops of water.

## 1.2 Amphoteric surfaces

The measurement of water contact angles ($\theta_w$) on typical amphoteric materials such as proteins is more complicated, because upon air-drying in a flat layer the surface properties of even a sturdy globular protein such as human serum albumin (HSA) tend to change from hydrophilic to hydrophobic. On the other hand, when one allows HSA to remain in aqueous solution (e.g., in a concentrated state on top of an ultrafilter membrane; van Oss and Good, 1988), one has to cope with hydration states that are difficult to estimate quantitatively. However, such measurements as exist (van Oss, Good and Chaudhury, 1986a; van Oss and Good, 1988; van Oss, 1989b) allow one to come to the following conclusions:

With $\theta_w$ measured at a pH below 4.8, HSA is positively charged and hydrophilic. Around its isoelectric point of $\approx$4.8, HSA has, by definition, a zero charge and is hydrophobic at that pH. This was measured with dried HSA which at pH 7 has a $\theta_w$ of 63.5° (van Oss and Good, 1988) and a $\theta_w$ of 75° at pH 4.8 (van Oss, 1989b). At pH values higher than 4.8 and further into the alkaline region, where HSA is more and more negatively charged and increasingly hydrophilic: at pH 7, hydrated HSA shows a water contact angle of 12.8° (van Oss, Good and Chaudhury, 1986b).

## 1.3 Importance of the $\zeta$-potential of simply electrically charged as well as of amphoteric surfaces

It should be understood that the influence of the pH of the water of contact angle drops on simply electrically charged as well as on amphoteric surfaces only acts because the pH of the water directly influences the $\zeta$-potential of these surfaces. Furthermore, changes in $\zeta$-potential concomitantly cause changes in the Lewis acid–base (AB) properties of such surfaces. See Chapter 3, Section 7, for a discussion on the $\Delta^{GEL}$–$\Delta G^{AB}$ linkage and see Chapter 8, Section 3 and in particular Sub-section 3.4, where the influence is discussed of added salts with plurivalent counterions on the surface tension properties of charged particles when immersed in water. Finally, see also the present chapter, Sub-section 3.3.

To summarize, the pH of the water used for contact angle measurements on simply electrically charged surfaces will influence their degree of hydrophilicity. For instance, with negatively charged surfaces such as clean glass surfaces which are negatively charged at all pH values when in contact with water, the higher the pH of the water, the more negatively charged and concomitantly, the more hydrophilic such surfaces become.

The pH of the water used for contact angle measurements on amphoteric surfaces (e.g., protein surfaces) influences the *sign of charge* as well as the *hydrophilicity/hydrophobicity* of such surfaces: A protein such as HSA, at low pH is positively charged and hydrophilic (low $\theta_w$); at its isoelectric pH of 4.8, HSA is uncharged and hydrophobic (elevated $\theta_w$) and at higher pH values than 4.8 HSA is negatively charged and once more hydrophilic (low $\theta_w$).

## 2. Influence of the pH on Water Contact Angles Measured on (Non-Charged) Hydrophobic as well as on (Non-Charged) Hydrophilic Surfaces

Water of the same pH values as those used on clean glass surfaces (see Section 1, above), i.e., pH 1.94, 6.1 and 12.8, was also used on dried dextran films (i.e., on neutral, hydrophilic surfaces), as well as on hydrophobic parafilm (van Oss and Giese, 2005).

On hydrophilic dextran the $\theta_w$ values at the three different pH's were: 34.8°, 35.8° and 36.7°, where none differed from the others to a statistically significant degree. Similar results were obtained with the exceedingly hydrophobic parafilm sheets. Here the $\theta_w$ values at the three different pH values were: 102.63°, 102.50° and 102.25°, between which no statistically significant differences could be identified (van Oss and Giese, 2005).

The total neutrality of the dextran had been ascertained earlier via moving boundary Tiselius electrophoresis by van Oss, Fike et al. (1974). At neutral pH and an ionic strength of 0.15 an exceedingly low $\zeta$-potential for dextran of −0.04 mV was derived from an observed electrophoretic mobility of $-0.002\ \mu m\, s^{-1}\, V^{-1}\, cm$. For parafilm on the other hand no $\zeta$-potential was measured and no published $\zeta$-potentials were found in the literature. However the electroneutrality of parafilm has nonetheless been assumed, not only because of its sole consistency of paraffins, with a complete absence of charged moieties in its chemical composition, but also because of the remarkable constancy of all three contact angles measured with water at the three widely differing pH values.

To summarize, on electrically neutral surfaces, be they hydrophilic (dextran) or hydrophobic (parafilm), the pH of the water used for contact angle measurements was shown to have no influence on the $\theta_w$ values obtained.

# 3. Influence of the Ionic Strength on Water Contact Angles

## 3.1 Low ionic strengths

At 1% (w/v) of an added salt such as NaCl, the surface tension of water ($\gamma_w$) increases by about 0.3 mJ/m$^2$. Thus, when using phosphate-buffered saline (PBS) (=0.89% NaCl, or $\mu = 0.15$) $\gamma_w$ increases by about 0.4% (Weast, 1970/71, p. F-28), which is not greater than the normal statistical fluctuations involved in the measurement of contact angles with water. Therefore, for $\theta_w$ measurements on layers of mammalian leukocytes or other blood cells, where maintaining physiological isotonicity is essential, the use of PBS is still the preferred approach (see van Oss et al., 1975).

## 3.2 High ionic strengths

At ionic strengths greater than $\mu = 1.0$, one arrives at a treatment called "salting-out." With salts such as $Na_2SO_4$ an aqueous solution of 1 M would increase the surface tension of water by about 4% (Weast, 1970/71, p. F-28) and thus at close to its saturation point, $Na_2SO_4$ would cause a surface tension increase of about 9%. This would translate in an increase in the free energy of cohesion of water of about 9% and thus would increase the hydrophobizing capacity of water to the same extent (9%). Saturated $(NH_4)_2SO_4$ would achieve an increase in the hydrophobizing capacity of water by about 13%. These are significant increases in the hydrophobizing capacity of water so that the use of such salts at high concentrations will "salt out" many proteins (i.e., cause them to precipitate), especially at pH values close to their isoelectric point. See Chapter 3, Section 6 and Chapter 5, Subsection 2.2.1.

## 3.3 Low concentrations of salts with plurivalent counterions

The hydrophobizing influence of adding small amounts of salts with, e.g., $Ca^{2+}$ or $La^{3+}$ cations to negatively charged solid particles is extremely pronounced; see Chapter 8, Section 3; see also this chapter, Table 12.1.

**Table 12.1** Summarizing the influence of the pH and of the presence of salts, at low and high concentrations, as well as of salts with plurivalent counterions, on the surface properties of electrically charged, amphoteric, or electrically neutral solids

| pH | Low salt (<0.2 M) | High salt (⩾1 M) | Plurivalent counterions (≪0.1 M) | Solid surface |
|---|---|---|---|---|
| + | 0 | +++ | +++[b] | Electrically charged |
| +++[a] | 0 | +++[a] | +++[b,c] | Amphoteric |
| 0 | 0 | (+)[d] | 0 | Uncharged |

a Strongest at pH closest to the isoelectric point of the amphoteric surface.

b Low concentrations of plurivalent counterions make oppositely charged surfaces hydrophobic and concomitantly also decrease their $\zeta$-potential.

c Plurivalent counterions only interact with amphoteric surfaces at pH values away from their isoelectric pH.

d Should be possible but I am unaware of experimental data confirming this.

# 4. Comparison Between the Influence of pH and Increases in Ionic Strength on Water Contact Angles on Solid Surfaces as well as on the Surface Properties of such Solid Surfaces when Completely Immersed in Water

The effects of using different pH values as well as of the admixture of different quantities and different sorts of salts on water contact angles measured on solid surfaces, compared with the effects of such treatments on the surface properties of solid surfaces when completely immersed in similarly treated water, are of course the same.

## 4.1 Uncharged solids or solid particles

Neither pH differences nor salts added to water have any noticeable influence on the surface properties of uncharged solids or solid particles or macromolecules exposed to such aqueous solvents. A minor exception might be the case of uncharged solids or solid particles or macromolecules exposed to a multi-molar aqueous salt solution, in which case uncharged particles might become "salted-out," e.g., an aqueous dextran or starch solution might precipitate or undergo a phase separation.

## 4.2 Influence of pH or added salt on electrically charged solid surfaces, particles or macromolecules

### 4.2.1 Regular electrically charged surfaces

The surface properties of regular electrically charged surfaces are influenced by changes in pH as well as by the mixture of salts, when exposed to or immersed in water; see Table 12.1.

### 4.2.2 Amphoteric surfaces

With amphoteric surfaces the admixture of high salt concentrations with the aim at insolubilization, is most effective *at or close to their isoelectric pH*. However, by their very nature, the hydrophobicity vs hydrophilicity status of amphoteric surfaces is the most readily influenced by either pH changes or by the admixture of added salts and in particular by the addition of low concentrations of plurivalent counterions, at the appropriate pH. Here, plurivalent counterion-driven destabilization works best at pH values *away from their isoelectric pH*. As a hydrophobizing mechanism, addition of plurivalent counterions is more effective than pH alone and in contrast with pH manipulations, treatment with plurivalent counterions is irreversible under normal conditions, although it can be reversed with complexing agents such as EDTA (see Wu et al., 1994a, 1994b).

## 5. Conclusions

Referring to Table 12.1, it is only completely uncharged materials whose surface properties are not influenced by either changes in pH or by the admixture of salts (with the possible exception of the administration of high salt concentration, which could conceivably be effective in "salting-out" some types of neutral macromolecules or particles).

The lack of influence of pH on electrically uncharged surfaces includes hydrophilic as well as hydrophobic surfaces, although entities with hydrophobic surfaces cannot be insolubilized in aqueous systems, *because they already are insoluble.* Also, aqueous solutions with low salt concentrations, of 0.2 M or less, will not noticeably influence the surface properties of, for example, living mammalian cells.

On the other hand, measurements of contact angles with water of different pH values and/or with water containing lots of salt, will significantly influence the surface properties of all regularly electrically charged as well as of amphoteric surfaces and should under virtually all circumstances be avoided. The same holds true for the use of water containing low concentrations of plurivalent counterions vis-à-vis electrically charged or amphoteric entities. Finally, it should be stressed once more that pH changes of water are especially hazardous to the integrity of the original surface properties of amphoteric materials.

CHAPTER THIRTEEN

# Macroscopic and Microscopic Aspects of Repulsion *Versus* Attraction in Adsorption and Adhesion in Water

## Contents



*Interface Science and Technology*, Volume 16
ISSN 1573-4285, doi: 10.1016/S1573-4285(08)00213-5

# 1. Macroscopic-Scale Repulsion vs Microscopic-Scale Attraction in Water

## 1.1 The adsorption of human serum albumin (HSA) onto metal oxide particles immersed in water

### 1.1.1 HSA adsorption onto monosized, amorphous, spherical silica particles

In the interaction between human serum albumin (HSA) molecules dissolved in water, and silica particles ($SiO_2$) suspended in an HSA solution, it has been shown by Docoslis, Wu et al. (2001) that the overall *macroscopic-scale repulsion* between HSA molecules and silica particles, expressed in kT units, is ≈5.1 times greater than the *microscopic-scale attraction* between the HSA molecules and discrete cationic sites on the $SiO_2$ particles.

Typically, monosized, amorphous, spherical silica particles of 2.1 μm diameter, were immersed in an aqueous HSA solution, at neutral pH, where it could be determined from the surface properties of the HSA molecules (1) and the $SiO_2$ particles (2) that these two entities *repelled* one another *on a macroscopic level* in water (w) to the extent of $\Delta G_{1w2}^{TOT(mac)} = +118.9$ kT, at "contact," i.e., at $d = d_0 = 0.157$ nm (Docoslis, Wu et al., 2001); for the derivation of $d_0$, see Chapter 3, Sub-section 1.2. [The +118.9 kT consists of: −0.6 kT (LW contribution), +114.9 kT (AB contribution) and +4.6 kT (EL contribution), Docoslis, Wu et al., 2001.]

On the other hand, the *microscopic-scale, attractive* interaction energy, $\Delta G_{1w2}^{TOT(mic)} = -23.5$ kT, for the same system. This was derived from the equilibrium adsorption constant measured at zero time $[K_{eq}^{(t\rightarrow 0)}]$, determined by means of Langmuir isotherms in which the slope of the isotherm curve at zero concentration yields $K_{eq}$. The extrapolation of $K_{eq}$ to $t \rightarrow 0$ was done by constructing different Langmuir isotherms, each based on adsorption data obtained after different time durations and by subsequently extrapolating to $t \rightarrow 0$; see Docoslis, Rusinski et al. (2001); see also Sub-sections 2.2 and 3.5, below. [The −23.5 kT consists of: −0.29 kT (LW contribution), −20.85 kT (AB contribution) and −2.32 kT (EL contribution), Docoslis, Wu et al., 2001].

Now, at first sight, the simultaneous existence of a *repulsive* $\Delta G_{1w2}^{TOT(mac)} = +118.9$ kT and an *attractive* $\Delta G_{1w2}^{TOT(mic)} = -23.5$ kT (derived via the two different but nonetheless entirely valid pathways outlined above) might tempt one to suspect that an error has crept in somewhere, because when two entities, immersed in water, *repel* one another with the considerable free energy of +118.9 kT they surely would not be able to approach each other closely enough to engage at the same time in a mutual *attraction* to the more modest extent of −23.5 kT? In an ideal model, consisting of perfectly smooth, spherical proteins interacting with equally spherical and smooth silica particles, this would indeed be unlikely. However, human serum albumin (HSA) is not really spherical (although still commonly called "globular"), but actually "V"-shaped (He and Carter, 1992) and although our smooth, monosized and amorphous $SiO_2$ beads are quite spherical and smooth on a *macroscopic* scale, their surfaces are not really *microscopically* homogeneous, as their surfaces are interspersed with numerous discrete, incompletely neutralized $Si^{(+)}$ ions, even though the particles' surfaces still have an overall (negative) ζ-potential of

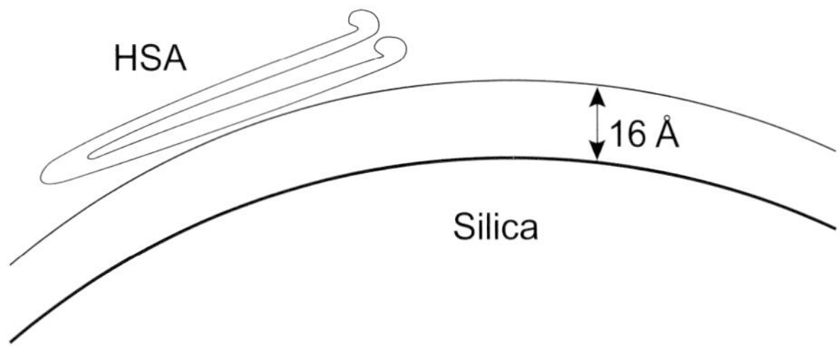


**Figure 13.1** The V-shaped human serum albumin (HSA) molecule approaching a silica sphere, oriented approximately *parallel* to the sphere's surface, in which orientation HSA and the silica sphere strongly repel each other, so that no closer approach than to about d $\approx$ 1.6 nm is possible; see Sub-section 1.1.1.

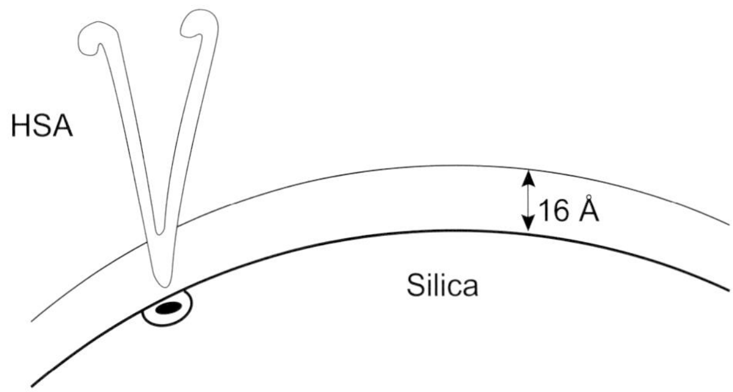


**Figure 13.2** The V-shaped human serum albumin (HSA) molecule approaching a silica sphere, oriented *perpendicular* to the sphere's surface, with the tip of the V situated just above and close to an $Si^{+}$ site (see He and Carter, 1992). Owing to the very small radius of curvature of the HSA molecule at the tip of its V, the HSA has been able to penetrate the repulsion field shown in Figure 13.1, and to bind quasi-specifically onto the silica sphere at the Si+ site; see Sub-section 1.1.1.

−56.2 mV (Docoslis et al., 1999). As far as the HSA molecules are concerned, their (also negatively charged) sharp point of the "V" can locally surmount the overall strong repulsion of +118.9 kT when that point is favorably oriented toward one of the $Si^{(+)}$ sites on the surface of a silica bead, because such a point of the "V" of an HSA molecule has a radius of curvature, r, of no more than 0.4 nm, whilst the average main radius, R, of the entire "V" of HSA is nearer to about 4 nm. This explains why *the point of the "V"* of HSA can readily penetrate the repulsion field of a silica bead thanks to its tiny radius r, whilst the HSA molecule in any other orientation [and thus approaching with an approximately ten times greater average radius R] is unable to overcome that repulsion (see Chapter 3, Section 8). This is schematically illustrated in Figures 13.1 and 13.2.

In Figure 13.1, with the "V" of an HSA molecule oriented *parallel* to the surface of a silica bead, it may be assumed that at $\Delta G^{TOT} \geqslant +10$ kT a limit is reached above which significant repulsion is assured (Overbeek, 1977), in which situation an HSA molecule cannot through ordinary spatial fluctuations approach the silica sphere's surface more closely than to d ≈ 1.6 nm (=16 Å), (calculated using the bottom equations of Tables 3.1–3.3).

In Figure 13.2, showing an HSA molecule approaching a silica sphere *perpendicular* to its surface, the tip of the "V" of HSA points to the sphere's center, just above and initially at d ≈ 16 Å removed from an $Si^{(+)}$ site. In this configuration and at d = 16 Å, the HSA molecule is attracted to the $Si^{(+)}$ site in question with a net $\Delta G_{1w2}^{TOT(mic)}$ of −2.0 kT. From here the HSA molecule can penetrate the repulsion barrier and proceed to $d = d_0$, helped by the fact that at short distances (i.e., at d < 16 Å), $\Delta G_{1w2}^{EL(mic)}$ can play an additional role in further in electrostatically attracting the HSA molecule to the silica surface by adding another -1.0 kT. This helps to direct an HSA molecule, with its tip of the "V" pointing to the silica surface, to be guided toward the nearest $Si^{(+)}$ site with a total of −2.0 kT − 1.0 kT = −3.0 kT from a distance of ≈16 Å, where it then finally proceeds to bind to it solidly with a $\Delta G_{1w2}^{TOT(mic)}$ of −23.5 kT at $d = d_0$ (Docoslis, Wu et al., 2001). In these conditions $\Delta G_{1w2}^{TOT(mac)} > 0$ and $\Delta G_{1w2}^{TOT(mic)} < 0$ can co-exist in the same system, accommodating a macroscopic-scale *repulsion* between immersed bodies with a large radius, R, as well as a microscopic-scale *attraction* involving a site with a small radius, r, which site is distally equipped with a ligand that can specifically bind to one of many receptors on the surface of one of the relatively much larger silica spheres.

### 1.1.2 HSA adsorption onto monosized carboxylated silica particles

The results outlined above, which were obtained with HSA and normal, monosized silica particles, may be compared with those obtained with HSA and similar monosized silica particles which were however carboxylated (here alluded to as monosized $COO^-$-silica particles) making them more strongly negatively charged. Their *repulsive* $\Delta G_{1w2}^{TOT(mac)}$ amounted to +151.2 kT, consisting of: −5.1 kT (LW contribution); +130.0 kT (AB contribution) and +26.3 kT (EL contribution), counterbalanced by $\Delta G_{1w2}^{TOT(mic)} = -17.7$ kT, consisting of: −2.5 kT (LW contribution); −13.7 kT (AB contribution) and −1.5 kT (EL contribution). Here the added $COO^-$ groups on the silica particles caused a significant increase in the macroscopic-scale repulsion of the HSA molecules by the silica particles, concomitantly with a decrease, but not a total cancellation of the microscopic-scale attraction between these two (Docoslis, Wu et al., 2001).

### 1.1.3 HSA adsorption onto polysized quartz particles

Next to the *monosized* amorphous silica particles discussed above (Sub-sections 1.1.1 and 1.1.2), HSA adsorption measurements were also done with polysized, multi-sharp-edged quartz particles, with an average particle size of 18.6 μm (Docoslis, Wu et al., 1999). With these more crystalline, edgy and non-spherical $SiO_2$ particles, results were obtained that were quite close to those achieved with the regular, smooth, monosized, amorphous, spherical silica particles: here $\Delta G1w2^{TOT(mac)}$ = +113.8 kT, consisting of: −0.6 kT (LW contribution), +107.8 kT (AB contribution) and +6.6 kT (EL contribution), whilst $\Delta G_{1w2}^{TOT(mic)} = -21.7$ kT, consisting of −0.27 kT (LW contribution) −20.32 kT (AB contribution) and −1.11 kT (EL contribution), Docoslis, Wu et al. (2001).

### 1.1.4 Comparison of HSA adsorption behavior onto different types of silica particles

While with the *regular*, monosized, smooth, hydrophilic silica spheres the ratio of $\Delta G_{1w2}^{TOT(mac)}/\Delta G_{1w2}^{TOT(mic)}$ (expressed in units kT) equals 5.07, with the polysized, multi-sharp-edged, hydrophilic quartz particles the $\Delta G_{1w2}^{TOT(mac)}/\Delta G_{1w2}^{TOT(mic)}$ ratio was 5.27. These ratios were virtually the same, indicating that the similar chemical composition and not the smoothness or otherwise of the two kinds of silica surfaces is the most important criterion in the quantitative interaction with HSA molecules. On the other hand, with the chemically altered, monosized, hydrophilic $COO^-$ silica beads the $\Delta G_{1w2}^{TOT(mac)}/\Delta G_{1w2}^{TOT(mic)}$ ratio had increased to 8.52, as a result of the stronger macroscopic-scale repulsion due to the increased electrostatic interaction energy caused by the added carboxyl groups on the particles' surfaces.

### 1.1.5 HSA adsorption onto hydrophobic talc particles

Finally, in addition to the adsorption of HSA molecules onto all these hydrophilic silica particles, the adsorption of HSA onto hydrophobic talc particles was also measured. Here, the $\Delta G_{1w2}^{TOT(mac)}$ repulsion was +39.1 kT, which is a much smaller (but still quite respectable) net repulsion energy than was found with any of the hydrophilic silica particles and was mainly due to the influence of the hydrophilicity of HSA, interacting with the (more modest) hydrophobicity of talc. $\Delta G_{1w2}^{TOT(mac)}$ consisted of: −0.4 kT (LW contribution), +35.8 kT (AB contribution) and +3.7 kT (EL contribution. $\Delta G_{1w2}^{TOT(mic)} = -24.8$ kT, consisting of −0.2 kT (LW contribution), −23.8 kT (AB contribution) and −0.8 kT (EL contribution) (Docoslis, Wu et al., 2001). Here one notes especially the more than three times smaller $\Delta G_{1w2}^{AB(mac)}$ with the hydrophobic talc particles, than with the hydrophilic silica ones, as talc is much less polar than silica. The total $\Delta G_{1w2}^{TOT(mic)}$ for talc, however, is of the same order of magnitude as with silica, which is understandable, as the *microscopic scale* attraction between HSA and silica and between HSA and talc follow essentially the same mechanism, as described above. Thus, with talc the ratio of macroscopic repulsion to microscopic attraction is not of the order of 5.1 or 8.5 as with HSA and silica, but only 1.58, which is mainly due to the much smaller repulsion between hydrophilic HSA and *hydrophobic talc* particles than between hydrophilic HSA and any of the equally hydrophilic silica particles.

## 2. Methodologies Used in Measuring Protein Adsorption onto and Desorption from Metal Oxide Particles in Water

### 2.1 The continuous circulation device

Given the necessity of maintaining the adsorbing particles in an unhindered suspension, a device was constructed which used a closed-circuit aqueous circulation of the (in this case HSA) solution which passed through a (magnetically stirred)

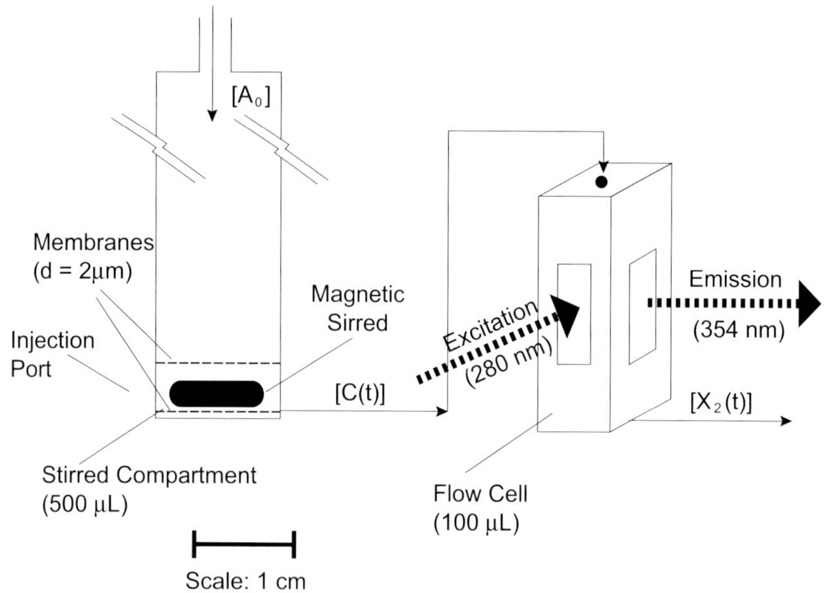


**Figure 13.3** Schematic diagram of the adsorption/desorption device mentioned in Sub-section 2.1. The adsorption (or desorption) takes place inside the magnetically stirred cylindrical flow chamber depicted at left. From the beginning of the experiment an aqueous protein solution circulates continuously through the entire apparatus and the protein concentration is measured every 0.1 s upon passing through the quartz flow cell (at right), by UV fluorescence spectrometry. Once the particles, whose surfaces are the adsorbing substrata, are quickly injected between the two membranes, into the stirred chamber between the two microporous membranes (of 2 μm pore diameter) situated at the lower part of the flow chamber, i.e., into the stirred compartment (already containing the flowing protein solution), a decrease in protein concentration is registered. This is because upon passing the stirred particle suspension, some of the protein adsorbs onto the particles' surfaces, thus causing an immediate *decrease* in the overall protein concentration, which is indicative of the amount of protein adsorbed. Desorption of protein, caused by the injection of a desorbing solute into the protein solution/particle mixture is measured in the same manner, via a registered sudden *increase* in protein concentration in the circulating solution. From Docoslis, Wu et al. (1999), with permission.

compartment of suspended metal oxide particles, enclosed between two microporous (Nuclepore) membranes with multiple monosized pores (of ⩽2 μm pore diameter) which are just small enough to prevent the particles from escaping; see Figure 13.3. After leaving the stirred compartment, the protein solution passes through a quartz flow cell equipped with a fluorescence spectrometer which continuously (i.e., every 0.1 s) measures the concentration of the passing protein solution, by using a monochromatic excitation wavelength of 280 nm, causing a fluorescent emission at 354 nm when detecting protein. This device served to measure protein adsorption (indirectly) onto metal oxide particles by analyzing the *decrease* in dissolved protein after passage through the particle compartment and it could also serve to measure protein desorption (directly) by analyzing the *increase* in protein concentration after having passed the particle compartment (Docoslis et al., 1999).

The results thus obtained served to obtain $K_{eq}^{t\to 0}$ as well as the kinetic association rate constant ($k_a$) and in some cases the kinetic dissociation rate constant ($k_d$); see Section 4, below and for the derivation of $K_{eq}^{t\to 0}$ see Sub-section 3.6, below.

## 2.2 Determination of $\Delta G_{1w2}^{(mac)}$ and $\Delta G_{1w2}^{(mic)}$

The macroscopic-scale $\Delta G_{1w2}^{(mac)}$ and the microscopic-scale $\Delta G_{1w2}^{(mic)}$ values, in the case of protein (and in this instance, HSA) adsorption onto metal oxide particles are determined via two entirely different approaches.

$\Delta G_{1w2}^{(mac)}$, which in all these instances is the free energy of (usually) *repulsion* between dissolved HSA molecules and the surfaces of, e.g., silica particles, so that here $\Delta G_{1w2}^{(mac)}$ has a positive value; see Figure 13.1. Now, free energies of repulsion are exceedingly difficult to measure in these, and probably also in most other conceivable cases. However, $\Delta G_{1w2}^{(mac)}$ *can* be derived from the respective surface properties of HSA (van Oss, 1994, 2006) and of the metal oxide particles in question (see, e.g., Docoslis, Rusinski et al., 2001), using Eq. (6.1) (see Chapter 6).

The determination of $\Delta G_{1w2}^{(mic)}$, which is the free energy of (usually) *attraction* between the point of the "V" of dissolved HSA molecules and cationic $Si^{(+)}$ sites on the surface of, e.g., silica particles (see Figure 13.2), cannot readily be measured directly via the surface properties of HSA and of the silica particles because this term is, by definition, based upon the *microscopic-scale* properties of these tiny interacting sites, which are too small for contact angle measurements. However, the value of $\Delta G_{1w2}^{(mic)}$ *can* be obtained via the experimentally determined value of the equilibrium association constant pertaining to the adsorption of HSA onto silica, i.e., $K_{eq}$, which can be directly derived from Langmuir adsorption isotherms, where the cosine of the slope of the isotherm curve at the point of zero HSA concentration yields $K_{eq}$ (Docoslis, Rusinski et al., 2001). Then, applying:

$$\Delta G_{1w2}^{(mic)} = [-kT \cdot \ln(K_{eq}^{t\to 0} \times 55.6)]/S_c \qquad (13.1)$$

(van Oss, Docoslis and Giese, 2001). The factor 55.6 (=the number of moles $H_2O$/L) is needed to transform L/mole into the dimensionless mole fractions. For the determination of $K_{eq}^{t\to 0}$, see Sub-section 3.7, below.

Many of the values for $\Delta G_{1w2}^{(mac)}$ and $\Delta G_{1w2}^{(mic)}$ which have been determined for various HSA/metal oxide particle interactions have been mentioned in Sub-section 1.1, above; see also Docoslis, Wu et al. (2001).

## 2.3 Influence of the pH of the aqueous medium on protein adsorption and desorption

### 2.3.1 Adverse results are caused by the use of *buffered* aqueous media

Docoslis, Rusinski et al. (2001) observed that phosphate buffered saline (PBS) would (at 21 °C) desorb about 67% of the HSA that had initially been adsorbed onto silica particles, compared with only 12% with pure $H_2O$, which was not due to the 0.89% NaCl content of the PBS, because just 0.89% NaCl-containing $H_2O$ did not desorb more HSA than pure $H_2O$. Furthermore, at 38 °C, pure $H_2O$ desorbed 42% of the initially adsorbed HSA, whilst PBS, also at 38 °C, desorbed about 96% of the adsorbed HSA; see also Rekveld (1997). It was therefore clearly the phosphate content of PBS that was the culprit. It turned out that all metal ion complexing agents appeared to act in this manner (e.g., phosphates, citrates, EDTA, etc.; see Docoslis, Rusinski et al., 2001). Of these, phosphates and citrates are among the more common anionic ingredients used in aqueous buffers in biochemical systems.

We therefore henceforth abstained from any further use of buffers in our protein adsorption and desorption experiments and we adjusted the pH of all aqueous media used in these protein adsorption measurements by just adding small amounts of HCl or NaOH, as needed. The only residual buffering effect we noted was the one provided by the presence of HSA.

We stress that only in the absence of all metal ion complexing compounds when measuring equilibrium as well as kinetic affinity constants in protein adsorption experiments, does one obtain unadulterated adsorption constants whose values are not compromised by the presence of chemicals that strongly compete for adsorption sites.

### 2.3.2 Influence of pH differences on the amphoteric properties of HSA

Close to the isoelectric pH ($\approx$4.8) of HSA, the pH of its aqueous solution is hard to control, because under these conditions HSA, now being at zero charge (and hydrophobic), can no longer contribute its own relatively modest buffering capacity.

However, *below* the isoelectric pH of HSA, i.e., at pH 3.0, when HSA is alkaline, unbuffered $H_2O$ desorbs only 4% of the HSA that had been adsorbed to silica beads. This is logical because under these conditions HSA is alkaline and $SiO_2$ is still mildly acidic. *Above* its isoelectric point, i.e., at pH 6.3 (when HSA has also turned acidic), about 7.5% of the adsorbed HSA desorbed from the silica and the amount *desorbed* increased gradually with increases in pH, to reach 97% at pH 10.5. On the other hand, also at pH 10.5, HSA would not adsorb at all onto any of the different types of silica surfaces discussed in Section 1, above (Docoslis, Rusinski et al., 2001; van Oss, Docoslis and Giese, 2001). The latter phenomenon was undoubtedly a consequence of the higher negative charge on HSA molecules, as well as on the silica particles at pH 10.5, so that the value of $\Delta G_{1w2}^{EL(mac)}$ became more positive (i.e., repulsive), as did $\Delta G_{1w2}^{AB(mac)}$ which is closely linked to $\Delta G_{1w2}^{EL(mac)}$ (see the "EL–AB linkage," discussed in Chapter 3, Section 7). It was therefore the enhanced mutual repulsion between HSA and silica at pH 10.5, that altogether prevented any HSA adsorption whatsoever, which is not surprising, as once adsorbed, HSA would have been for 97% desorbed at that pH anyway (Docoslis, Rusinski et al., 2001).

## 2.4 Other desorption approaches

Apart from changes in pH, there is the influence of time, then there is the influence of increases in temperature, and of the admixture of a number of different solutes which can enhance the desorption of protein (e.g., HSA) from metal oxide surfaces, such as silica as well as talc.

### 2.4.1 The influence of time on protein desorption

The longer the initial duration of the protein adsorption process, the smaller the amount of protein that can be desorbed in pure water as well as in PBS. This general phenomenon is a consequence of the *hysteresis* of adsorption, i.e., of the strengthening with time of the free energy of attachment between protein and the metal oxide surface; see Docoslis, Rusinski et al. (2001) and Section 3, on hysteresis, below.

### 2.4.2 The influence of temperature on protein desorption

Most of the work cited in this chapter involved the study of effects caused by differences in temperature (T) on the adsorption of HSA onto metal oxide particles used only 21 and 38 °C. This limit was imposed because a further increase in T, e.g., to about 56 °C or above, could begin to cause a denaturation in the protein molecule. In any event, in pure water, as well as in PBS the desorption of HSA from silica surfaces is stronger at 38° than at 21 °C: At neutral pH the desorption with $H_2O$ is about 13% at 21 °C and 42% at 38 °C, whilst with PBS it is about 69% at 21 °C and 96% at 38 °C (Docoslis, Rusinski et al., 2001). The latter was solely due to the *phosphate* in the PBS, because just 0.15 M NaCl desorbed exactly as much HSA as pure water, at both temperatures.

### 2.4.3 The influence of various solutes on protein desorption

A comparison was made between the admixture of 30% glucose, and 30% dextran (a polymer of maltose, which is a dimer of glucose). At neutral pH (pH 6.9) both additives desorbed HSA almost to the same extent: Glucose: 7% and dextran 11%. However, at pH 9.8, 30% glucose still desorbed only about 6.5% whilst 30% dextran desorbed about 34% of the adsorbed HSA. Here the 30% dextran [which has only a slight (lowering) influence on the surface tension of water] behaved more like pure water at the two pH values used, whilst with 30% glucose, especially at pH 9.5, almost six times less HSA was desorbed than with pure water or with 30% dextran. Now, *30% glucose* significantly *increases* the surface tension of water (by about 32%, see Docoslis, Giese and van Oss, 2000) and thus also increases its free energy of cohesion and, concomitantly, its hydrophobizing capacity (see Chapter 5, Sub-section 2.2.1). Thus the 30% glucose would *strengthen* the adsorptive bond between HSA and silica through the latter's hydrophobization and in so doing prevent much of the desorption of HSA from silica (Docoslis, Rusinski et al., 2001).

One of the strongest desorbing agents turned out to be 4.5 M urea, which desorbs about 96% HSA at pH 7.4 and 100% at pH 9.6 (which is its natural pH at 4.5 M; Docoslis, Rusinski et al., 2001).

# 3. HYSTERESIS OF PROTEIN ADSORPTION ONTO METAL OXIDE SURFACES, IN WATER

## 3.1 Hysteresis' interference with the determination of $K_{eq}$ and $k_d$

The Greek word, "hysteresis" (delay, deficiency), when used in the context of adsorption phenomena signifies the strengthening of adsorptive bonds as a function of time. This bond-strengthening with time starts in most cases within a fraction of a second from the very moment the bond-forming contact has been made. In the case of practically all adsorptive bonds (with the exception of certain purely electrostatic bonds, see Sub-section 3.4, below), this means that in the cases of $K_{eq}$ and $k_d$, where pre-adsorption is a pre-requisite for their experimental determination, one starts by creating severe hysteresis before even doing the first measurements, thus precluding their direct experimental measurement. With the determination of $K_{eq}$ one can circumvent this problem by extrapolation to zero time; see Sub-section 3.6, below. However in the case of the kinetic dissociation rate constant, $k_d$, it is not practicable to circumvent the hysteresis problem in this manner, so that $k_d$ must be derived via a different approach; see Sub-section 4.4.5, below.

## 3.2 Hysteresis following hydrophilic adsorption

In the case of adsorption where both ligands (e.g., HSA) and receptors (e.g., $Si^{(+)}$ sites on silica surfaces) are hydrophilic, a $\Delta G_{1w2}{}^{(mic)}$ adsorptive bond can, upon extrusion of interstitial water after its initial formation, change from $\Delta G_{1w2}{}^{(mic)}$ into a $\Delta G_{12}{}^{(mic)}$ type of bond, which strengthens the bond considerably (see also Israelachvili and Wennerstrőm, 1996). This is because $|\Delta G_{1w2}|$ is smaller than $|\Delta G_{12}|$ (i.e., after expulsion of interstitial water). The reason for this is the following:
given that:

$$\Delta G_{1w2} = \gamma_{12} - \gamma_{1w} - \gamma_{2w} \tag{13.2}$$

[see Dupré equation (2.12)], and:

$$\Delta G_{12} = \gamma_{12} - \gamma_{1} - \gamma_{2} \tag{13.3}$$

[see Dupré equation (2.8)], with hydrophilic or weakly hydrophobic entities:

$$\gamma_i > \gamma_{iw}$$

so that in the case of HSA and silica (entities 1 and 2) the value for $\Delta G_{1w2}{}^{IF(mic)}$ was $-20.2$ mJ/m$^2$, as derived from the surface properties of HSA and silica, see van Oss, Docoslis and Giese (2001) [the superscript IF alludes to "interfacial", i.e., comprising LW + AB contributions only (not including EL forces)]. From the same (mic) surface data (van Oss, Docoslis and Giese, 2001) the (post-hysteresis) value for $\Delta G_{12}{}^{(mic)}$ was obtained, amounting to $-133.5$ mJ/m$^2$.

## 3.3 Hysteresis following adsorption onto a hydrophobic surface

Despite the macroscopic-scale hydrophobicity of talc ($\Delta G_{2w2}^{(mac)} = -48.7$ mJ/m$^2$), the strong macroscopic-scale hydrophilicity of HSA ($\Delta G_{1w1}^{(mac)} = +85.5$ mJ/m$^2$) is such that, macroscopically, HSA and talc still *repel* one another in water by $\Delta G_{1w2}^{(mac)} = +16.5$ mJ/m$^2$, based on surface properties data from van Oss, Docoslis and Giese (2001). Nonetheless, HSA *adsorbs* onto talc via a net microscopic-scale attraction of the same type as depicted in Figure 13.2. However, with talc the pathway followed by the enhanced attraction due to hysteresis is then a *hydrophobic* one, in contrast with the *hydrophilic* hysteretic pathway described for the HSA-silica case in the preceding Sub-section 3.2, above).

HSA initially binds to talc in the same manner by which HSA binds to silica, i.e., by means of the point of the "V" of HSA being attracted by a still partly ionized metal cation on the surface of the metal oxide particle. Secondary, hysteretic, hydrophobic bonds form subsequently as a consequence of the attraction between the general hydrophobic talc surface and the normally internalized hydrophobic sites of HSA which externalize under the influence of the vicinity of the relatively large hydrophobic talc particle, as earlier described in connection with the attraction between HSA and the exceedingly hydrophobic water-air interface, occurring during air-drying of aqueous HSA solutions; see Chapter 11, Sub-section 3.3.2. Taking the air-dried surface tension data for HSA and the macroscopic-scale surface tension data for talc (van Oss, Docoslis and Giese, 2001), the value for $\Delta G_{1w2}^{(hysteresis)} = -33.8$ mJ/m$^2$. This is the *additional* free energy of attraction between HSA and talc, immersed in water, due to hysteresis, where the HSA has been hydrophobized following its initial microscopic-scale attraction to talc. As the pre-hysteresis free energy of (microscopic-scale) attraction between HSA and talc; $\Delta G_{1w2}^{(mic)} = -48.9$ mJ/m$^2$ (from surface tension data given in van Oss, Docoslis and Giese, 2001), the maximum possible attraction energy between HSA and talc would amount to $\Delta G_{1w2}^{(mic)} + \Delta G_{1w2}^{(hysteresis)} = -33.8 - 44.9 = -83.7$ mJ/m$^2$, which may however never be entirely reached. [With this type of ($\Delta G_{1w2}^{(mic)} + \Delta G_{1w2}^{(hydrophobic)}$) adsorption, the two contributing energies are clearly of separate origins and therefore additive.]

## 3.4 Absence of hysteresis when adsorptive forces are purely electrostatic

In cases of auto-antibody formation in patients with systemic lupus erythematosus (SLE), one often finds auto-antibodies directed against double-stranded DNA (ds-DNA) in the patients' sera, which are called "nuclear antibodies, or "anti-nuclear antibodies". These antinuclear (anti-DNA) antibodies generally react with ds-DNA via purely electrostatic bonds. We found that in such patients' sera low or medium avidity anti-ds DNA antibodies would occur with which the energy involved in *preventing* the bond formation between anti-ds DNA antibodies and ds-DNA (using *Crithidia luciliae* kinetoplasts as the source of ds-DNA) was the same as the energy needed to *dissociate* the bond.

In other words, in these cases of purely electrostatic antigen–antibody bonds, no hysteresis occurred, or, as the conclusions of this work explained: "the electrostatic component of this type of antigen–antibody bond does not become stronger with time, for time spans up to one hour" (Smeenk et al., 1983; van Oss, Smeenk and Aarden, 1985). As in the normal course of events, with most other physical, non-covalent bond formations between different entities, when immersed in water, hysteresis sets in immediately, i.e., in less than a second, it seems safe to state that such electrostatic anti-ds-DNA/ds-DNA bonds do not undergo hysteresis. This absence of hysteresis appears to be due to the fact that purely electrostatic bonds between positively charged *paratopes* (i.e., antibody-active sites) and negatively charged epitopes (i.e., ds-DNA surfaces) do not undergo hydrophobic interactions because they are both hydrophilic and apparently do not undergo extrusion of interstitial water after bonding either, so that the purely electrostatic forces that are active in their bond formation require precisely *the same energy* to dissociate their bond *and not more*.

However, whilst with the *highest avidity* antisera prevention of association as well as dissociation were about the same, using high pH (≈pH 12), dissociation of high-avidity anti-ds-DNA/ds-DNA using high ionic strengths is not possible, even when using 5 M NaCl; see Smeenk et al. (1983). It was concluded in the 1983 paper (quoted above) that as far as purely electrostatic bonds are concerned, these undergo no hysteresis (within the first hour after the start of bond formation). However the fact that high pH (pH 12) *could* dissociate the highest avidity anti-ds-DNA/ds-DNA, whilst admixture of even 5 M NaCl *could not*, indicated to us at the time (cf. the 1983 and 1985 papers, quoted above) that the highest affinity anti-ds-DNA/ds-DNA bonds appeared to comprise strong hydrogen bonds as well as electrostatic bonds. This conclusion was arrived at because the experiments showed that these (strong) hydrogen bonds *could* be dissociated at high pH, but *not* in the presence of high salt Smeenk et al., 1983; van Oss, Smeenk and Aarden, 1985.

It should be remembered however that up to and including the middle 1980's neither the present author and his collaborators, nor anyone else, had yet arrived at a precise notion of the mechanism of hydrophobic interactions. From what we *now* understand however, the second, strong non-electrostatic anti-ds-DNA/ds-DNA bond of the highest avidity system was in reality more likely to have been a *hydrophobic interaction*. This is because a hydrophobic attraction would actually have been *strengthened* by the admixture of high concentrations of salt, as was indeed the case. See Chapter 5, Sub-section 2.2.1 and Chapter 6, Sub-section 2.4.2 for *the influence of salt* on hydrophobic interactions between two different compounds or materials, immersed in water. Thus this non-dissociation of a bond between the highest affinity anti-ds-DNA and ds-DNA by the admixture of lots of salt, in contrast with the successful dissociation of that bond at high pH, would be more aptly described as a "salting-out" effect.

The effect at pH 12 would be that the (initially *basic*) antibody-active site (i.e., the paratope; see Chapter 14, below) would become *acidic* and the already acidic ds-DNA (the antigen) would become even more acidic at high pH and both would, concomitantly, become more hydrophilic, so that they would mutually repel one another at high pH, both on account of an electrostatic repulsion and through their mutual hydrophilicity at that pH, both of which would cause their bond to dissociate.

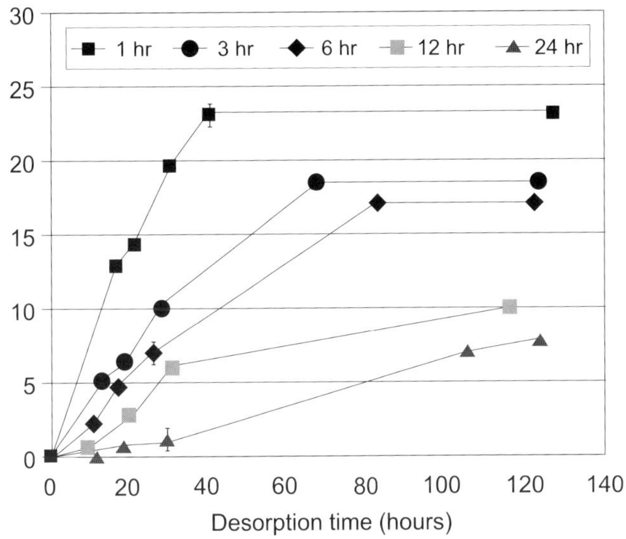


**Figure 13.4** Kinetics of the desorption of HSA from silica in pure water, at 21 °C and pH 7.4. For each initial adsorption time period, in particular from 1 to 6 h, a break occurs in each graph, beyond which given time lapse, irreversibility of the hysteretic binding sets in, which occurs at 38 h (1 h initial contact); 63 h (3 h initial contact) and 80 h (6 h initial contact); see Sub-section 3.5. From van Oss, Docoslis and Giese (2001), with permission; see also Docoslis, Rusinski et al. (2001).

The explanation of the lack of hysteresis of this purely EL system is given in Chapter 14, devoted to "Specific Interactions in Water," Sub-section 2.2.3; see below.

## 3.5 Hysteresis as a function of adsorption time

The pre-hysteresis equilibrium desorption constant ($K_d$) value at time zero and the pre-hysteresis value of the kinetic desorption rate constant ($k_d$) are not easily obtainable via direct measurements. This is because before one can measure dissociation equilibrium dissociation constants or dissociation rate constants one needs to pre-adsorb a sufficient amount of protein, which then can be desorbed afterwards, where the "afterwards" usually involves at least half an hour's worth of *adsorption* time before can start measuring equilibrium desorption constants or kinetic desorption rate constants. However, in Sub-sections 4.4 and 4.4.5, below, a method is given to estimate rather reliable $k_d$ values via a more indirect approach and in Sub-section 4.4.5 the ill-advisedness of the frequent use of the term "$K_d$" is discussed.

Meanwhile, Figure 13.4 illustrates hysteretic behavior via an analysis of the desorption of HSA molecules from silica surfaces as a function of the initial adsorption contact time (Docoslis, Rusinski et al., 2001; van Oss, Docoslis and Giese, 2001). The initial contact times during which HSA was being adsorbed from an aqueous solution at pH 7.4 and 21 °C, at 1, 3, 6, 12 and 24 h, after which in each case the amount of desorbed HSA was measured with just pure water, still at pH 7.4 and

21 °C. It is clear, especially from the relatively short initial contact periods, i.e., 1, 3 and 6 h; see Figure 13.4, that after a given desorption period in pure water and also at pH 7.4 (in these cases after approximately 38, 70 and 83 h, respectively), a break finally sets in, after which no further measurable desorption occurs. It is also clear from Figure 13.4 that the longer the initial exposure of the silica spheres to HSA, the smaller the amount of HSA that can be desorbed. The breaks show points occurring many hours after the initial exposure times of silica to HSA of 1, 3 and 6 h, at which points the curves become horizontal, indicating that to the right of these points no further measurable desorption occurs. This also indicates the *irreversibility* of the hysteresis, which is all the more pronounced (through the cessation of further desorbable HSA), the longer the duration of the initial exposure to HSA. It is also feasible, using the data depicted in Figure 13.4, to deduct that by extrapolating to an initial contact time of, e.g., one minute, only about 28% of the initially adsorbed HSA would be desorbed from the silica, leaving 72% irreversibly adsorbed.

It should finally be noted that serum albumin, while remaining irreversibly adsorbed onto a metal oxide surface such as a glass plate, nonetheless still retains the freedom of unimpeded lateral movement along the glass surface, when being driven by diffusional (Michaeli et al., 1980) or by electrophoretic forces (Absolom et al., 1981a, 1981b).

## 3.6 Importance of using the value for pre-hysteresis $K_{eq}^{t\rightarrow 0}$

Using the correct value for the equilibrium adsorption constant, $K_{eq}$, is important, because it is the basis for deriving the value of $\Delta G_{1w2}^{(mic)}$; see Eq. (13.1), in Sub-section 2.2, above. Together with the kinetic *adsorption rate* constant, $k_a$, knowledge of the accurate value of $K_{eq}$ (in the form of $K_{eq}^{t\rightarrow 0}$) is also a prerequisite for obtaining a reliable value for the kinetic *desorption* rate constant, $k_d$; see Sub-section 4.4.5, below. (For the determination of $k_a$, see Sub-section 4.3, below.)

It should be noted, meanwhile, that there is no need to designate $K_{eq}^{t\rightarrow 0}$ as "$K_{eq}^{(mic)}$" because there can be no "$K_{eq}^{(mac)}$" that could be linked to a $\Delta G_{1w2}^{(mac)}$ because the latter is generally *repulsive* and there is no actual attractive bond for which one might be able to express an equilibrium adsorption constant as "$K_{eq}^{(mac)}$." Thus, $K_{eq}^{t\rightarrow 0}$ is the equilibrium binding constant pertaining *exclusively* to *microscopic-scale* adsorption (here of HSA) and hence is only connected to $\Delta G_{1w2}^{(mic)}$; see Eq. (13.1).

## 3.7 Determination of $K_{eq}^{t\rightarrow 0}$

$K_{eq}$ is most conveniently measured as follows: In using the Langmuir isotherm approach, one plots the fractional surface coverage of protein through adsorption (on the ordinate) vs. the different dissolved protein concentrations used (on the abscissa), after which one measures the angle of the tangent at the zero point of the isotherm curve formed with the abscissa, which then yields $K_{eq}$. Unfortunately, in the normal course of measuring $K_{eq}$ applied to a case of protein adsorption, once one starts the measurements required for constituting a Langmuir isotherm comprising the amount of protein adsorbed vs. the protein concentration, the very moment one adds the first dissolved protein, *hysteresis sets in*. This poses a serious problem because one cannot avoid giving the dissolved protein some time to adsorb onto the substratum before one can measure the amount of protein that has been adsorbed, i.e., typically one or more *hours*, after which a serious amount of hysteresis has occurred (see Figure 13.4 and see also Docoslis, Rusinski et al., 2001). This pre-measurement hysteresis will give rise to anywhere from an increase of 35% to many hundred % in the $K_{eq}$ values determined via the regular Langmuir isotherm approach, depending on the preliminary exposure time used for the initial adsorption step.

There is, however, a way to remedy this large hysteresis-driven inaccuracy in measuring $K_{eq}$ via commonly practiced Langmuir isotherm determinations and that is to construct such isotherms based on a number of different initial protein exposure times (e.g., 1, 3 and 6 h) and thence to extrapolate to *zero exposure time*, thus permitting the determination of $K_{eq}^{t\to 0}$; see Docoslis, Rusinski et al. (2001), where data are given for the value of $K_{eq}^{t\to 0}$, compared with $K_{eq}$ values obtained after 1, 3, 6. 12 and 24 h pre-adsorption. For example, taking the polysized silica and HSA system, $K_{eq}^{t\to 0}$ is 2.5 times smaller than for $K_{eq}$ determined after 1 h pre-adsorption, 6.6 times smaller than for $K_{eq}$ determined after 3 h pre-adsorption and 13 times smaller than for $K_{eq}$ determined after 6 h pre-adsorption.

## 4. Kinetics of Protein Adsorption onto Metal Oxide Surfaces Immersed in Water

Under ideal conditions, where only diffusion-driven adsorptive attachment occurs between protein molecules, dissolved in water and a solid surface, immersed in water, with neither attraction at, nor repulsion from a distance between the protein molecules and the solid surface and in the absence of flowing or stirring, the kinetic adsorption rate constant, $k_a$ can be described as follows (Docoslis et al., 1999; van Oss, 2006, Chapter XXV):

$$k_a = 4\pi d_0 D(N/1000), \tag{13.4}$$

where $k_a$ is expressed in $cm^3/(mM\,s)$ [which is equivalent to $L/(M\,s)$]; $d_0$ is the minimum equilibrium distance to which, e.g., an HSA molecule can approach a silica surface (see Chapter 3, Sub-section 1.2); D is the diffusion constant of the protein molecule when dissolved (at 20 °C, $D \approx 6.1 \times 10^{-7}$ $cm^2/s$, for HSA); and N is Avogadro's number ($6.02 \times 10^{23}$ molecules per Mol), making the ideal $k_a$ value of this interaction to amount to $7.2 \times 10^7$ L/M s.

However, adsorption conditions are never ideal, due to avoidable as well as unavoidable experimental constraints. Some of the *avoidable* experimental issues have been treated in part in Sub-section 3.7, above, concerning the determination of $K_{eq}^{t\to 0}$ and among some of the other *avoidable* undesirable experimental conditions is the determination of $k_a$; see Sub-section 4.3, below.

The influence of the main unavoidable experimental conditions has been treated by von Smoluchowski (1918), see Sub-section 4.1, below.

## 4.1 Von Smoluchowski's approach applied to the kinetic adsorption rate constant, $k_a$

In one of von Smoluchowski's last papers (he died in 1917) he described a treatment of the kinetics of flocculation of particles suspended in a liquid (von Smoluchowski (1917), but published in 1918), which treatment applies equally well to the adsorption of proteins, dissolved in water, onto metal oxide particles with discreet adsorptive sites imbedded in their surfaces, which also are immersed in water (van Oss, Docoslis and Giese, 2001; Docoslis, Wu et al., 1999, 2001; see also van Oss, 2006, Chapter XXV).

## 4.2 Von Smoluchowski's f factor

To take into account both the "improbability" of the occurrence of an interaction (in our case exemplified by a *repulsion*, and taking place because of an unfavorable orientation of the protein) as well as its "probability" (here represented by an *attraction*, made possible by a favorable orientation of the protein), von Smoluchowski (1918) introduced a factor, **f**, into Eq. (13.4):

$$k_a = 4\pi d_0 D\mathbf{f}(N/1000), \tag{13.5}$$

where:

$$\mathbf{f} = \int_{\phi} \exp[1/d \int_{d=d_0}^{d=\infty} \{(-\Delta G_{1w2}{}^{TOT})/kT\}\, dd]\, d\varphi \tag{13.6}$$

[here *d* and $d_0$ (for distance) are shown in *italics*]. In Eq. (13.6) $\phi$ stands for all *orientations* of the dissolved protein molecules, in such a manner that the favorable orientations (a) are related to the unfavorable orientation (1 − a) as:

$$\varphi = a/(1 - a). \tag{13.7}$$

With the help of Eqs. (13.5)–(13.7) it is possible to subdivide $k_a$ (which can be measured experimentally with a fair degree of accuracy; see Sub-section 4.3, below) into $k_a{}^{(mac)}$ and $k_a{}^{(mic)}$ where, however, $k_a{}^{(mac)}$, like $K_{eq}{}^{(mac)}$, has no meaning (see Sub-section 3.6, above). On the other hand, $k_a{}^{(mic)}$, in combination with $K_{eq}{}^{t\to 0}$, allows the determination of the kinetic *desorption* rate constant, $k_d{}^{(mic)}$, using:

$$k_d{}^{(mic)} = k_a{}^{(mic)}/K_{eq}{}^{t\to 0}. \tag{13.8A}$$

See also Eq. (13.8B) in Sub-section 4.4, below. The approach made possible by using Eq. (13.8A) is extremely useful because $k_d{}^{(mic)}$ cannot readily be measured experimentally because any such attempt requires a pre-adsorption step which causes a considerable amount of irreversible hysteresis; see Sub-sections 4.4 and 4.4.5, below.

## 4.3 Determination of f and $\varphi$

To determine the values for **f** and $\phi$, the following pathway has been used:

First the $\Delta G^{1w2}$ term of Eq. (13.6) has to be subdivided as follows:

$$\chi^{TOT} = 1/d \cdot \int_{d=d_0}^{d=\infty} [(-\Delta G_{1w2}{}^{TOT})/kT]\, dd, \tag{13.9}$$

where:

$$\chi^{TOT} = a \cdot \chi^{(mac)} + (1 - a)\, \chi^{(mic)} \tag{13.10}$$

and where:

$$\chi^{(mac)} = 1/d \cdot \int_{d=d_0}^{d=\infty} [(-\Delta G_{1w2}{}^{(mac)})/kT]\, dd \tag{13.11}$$

and:

$$\chi^{(mic)} = 1/d \cdot \int_{d=d_0}^{d=\infty} [(-\Delta G_{1w2}{}^{(mac)})/kT]\, dd \tag{13.12}$$

which makes it possible to rewrite von Smoluchowski's **f** factor as:

$$\mathbf{f} = -\exp \chi^{TOT} = -\exp[a \cdot \chi^{(mac)} + (1 - a) \cdot \chi^{(mic)}] \tag{13.13}$$

or:

$$\ln \mathbf{f} = \ln f^{(mac)} + \ln \mathbf{f}^{(mic)} \tag{13.14A}$$

so that:

$$\mathbf{f} = \mathbf{f}^{(mac)} \cdot \mathbf{f}^{(mic)}, \tag{13.14B}$$

where:

$$\mathbf{f}^{(mac)} = -\exp[a \cdot \chi^{(mac)}] \tag{13.15}$$

and:

$$\mathbf{f}^{(mic)} = -\exp[(1 - a) \cdot \chi^{(mic)}], \tag{13.16}$$

see Docoslis, Wu et al. (1999, 2001); see also van Oss (2006, Chapter XXV) and for the measurement of $\Delta G_{1w2}{}^{(mac)}$ and the derivation of $\Delta G_{1w2}{}^{(mic)}$ see Subsection 2.2, above. From these values, at $d = d_0$, one can obtain $\chi^{(mac)}$ and $\chi^{(mic)}$, using Eqs. (13.11) and (13.12). Then, still using Eqs. (13.11) and (13.12), one can utilize the extended DLVO (XDLVO) approach, described in Chapter 3, using Tables 3.1–3.3, applied to all distances, $d$, in practice up to $d = 10$ nm (except when using completely de-ionized water, where one may have to extend $d$ to at least 100 nm). With XDLVO plots thus obtained one derives the integrated parts (right-hand side of Eqs. (13.11) and (13.12) via the surface areas under the XDLVO curves, down to (or up to) the abscissa (as the case may require); see van Oss, Wu et al. (2001); van Oss, Docoslis et al. (2002).

The **f** factor can then be obtained from the measured $k_a$ value (Eq. (13.5)) where however the diffusion coefficient, D (in our case the diffusion coefficient of HSA), has to be multiplied by a factor (here 62.5) to account for the forced fluid flow around the immersed suspended particles (see Figure 13.3 and see also Docoslis, Wu et al., 1999), thus yielding $k_a^{(mic)}$.

As all three $\chi$ values can now be derived [Eq. (13.9) for $\chi^{TOT}$ and Eqs. (13.11) and (13.12) for $\chi^{(mac)}$ and $\chi^{(mic)}$] and using the above-described XDLVO approach for solving the applicable integrals, one can also obtain the values for a and $(1 - a)$ via Eq. (13.10), thus also yielding the $\varphi$-factor (Eq. (13.7)) as well as the value for **f** (using Eq. (13.13)). For our final results *re* $\chi^{(mic)}$, $\mathbf{f}^{(mic)}$, $k_a^{(mic)}$, $K_{eq}^{t\to 0}$ (and also $k_d^{(mic)}$, see Sub-section 4.4.5), for the HSA adsorption systems comprising: polysized, monosized and monosized-$COO^-$ silicas, as well as talc, see Docoslis, Wu et al. (2001, 2002), where the 2002 entry is an *erratum* pertaining to the 2001 paper.

## 4.4 The equilibrium binding constant and the kinetic rate constants

### 4.4.1 Pronounced variability of $k_d^{(mic)}$

Regarding the data pertaining to the interaction constant $K_{eq}^{t\to 0}$ and the kinetic $k_a^{(mic)}$ and $k_d^{(mic)}$ values for the HSA-monosized silica bead values on the one hand and the HSA-monosized silica-$COO^-$ values on the other hand, which are, respectively, (cf. Docoslis, Wu et al., 2001, 2002):

monosized silica ($SiO_2$): $K_{eq}^{t\to 0} = 2.78 \times 10^8$ L/M; $k_a^{(mic)} = 2.66 \times 10^8$ L/M s and $k_d^{(mic)} = 0.96/s$ ($\varphi = 1.364$).

vs:

monosized silica-$COO^-$ ($SiO_2$-$COO^-$): $K_{eq}^{t\to 0} = 4.2 \times 10^6$ L/M; $k_a^{(mic)} = 1.61 \times 10^8$ L/M s and $k_d^{(mic)} = 38.3/s$ ($\varphi = 0.848$).

it becomes apparent that the major difference between the standard monosized ($SiO_2$) and the much more negatively charged ($SiO_2$-$COO^-$) particles is found to lie in the $k_d^{(mic)}$ values which in the case of the monosized ($SiO_2$-$COO^-$) particles shows that HSA spontaneously *desorbs* from the latter about *40 times faster* than from the untreated ($SiO_2$) particles, even though HSA *adsorbs* at almost the same rate to the $SiO_2$ as to the $SiO_2$-$COO^-$ particles. The differences in $K_{eq}^{t\to 0}$ are mainly a consequence of the relation between $K_{eq}^{t\to 0}$ on the one hand and $k_a^{(mic)}$ and $k_d^{(mic)}$ on the other hand, see Sub-section 4.4, and Eq. (13.8B), below. Among the latter two kinetic rate constants it is clear that the large divergence between the two $K_{eq}^{t\to 0}$ values is almost entirely due to the large difference between the two $k_d^{(mic)}$ values (cf. Eq. (13.8A); see also Eq. (13.8B), below). This observation on the pronounced variability of $k_d^{(mic)}$ correlates closely with earlier findings by Absolom and van Oss (1986) that among 14 different hapten/antihapten systems, with the kinetic dissociation rate constant, $k_d$, the boundaries of variability were a factor

230,000 farther apart than was the case with those of $k_a$; see also Sub-section 4.4.4, and Chapter 14, below.

### 4.4.2 Experimental determination of $k_a$

The methodology described in Sub-section 4.4.1, above (see also Figure 13.3) was actually primarily intended for the measurement of $k_a$, for which it lends itself extremely well provided one observes the following two precautions which are both of critical importance (Docoslis, Wu et al., 1999):

1. $k_a$ needs to be measured within the shortest possible time (i.e., within a fraction of a second) after the addition of (here) dissolved protein, and:
2. $k_a$ should be measured at the lowest possible protein concentration.

**Ad 1** It has been shown that whilst a $k_a$ value measured within 0.1 s from the start, yielding $k_a = 3.7 \times 10^6$ L/M s, this value decreased to $k_a = 0.6 \times 10^6$ L/M s after 20 s, thus effectively *decreasing more than six-fold* after 20 s (all done at a protein concentration of 7.25 nM; Docoslis, Wu et al., 1999).

**Ad 2** It has been shown that whilst at a protein concentration extrapolated to zero nM the measured $k_a$ was $4.53 \times 10^6$ L/M s and at 7.25 nM $k_a$ was found to be $3.71 \times 10^6$ L/M s and at 35.7 nM it yielded $k_a = 1.23 \times 10^6$ L/M s, finally, at 14.5 μM protein concentration, $k_a$ appeared to be $0.14 \times 10^6$ L/M s. This amounts to a *decrease* in the $k_a$ value measured at the highest, compared to the lowest protein concentration used, *of a factor 32.4* (see Docoslis, Wu et al., 1999).

Based on the use of commercially available devices at the above cited dates, one encountered in the literature $k_a$ values measured after time lapses of the order of at least a minute and up to 20 min. Such measurements must have resulted in $k_a$ values that were two to three decimal orders of magnitude too low (van Oss, 1999).

### 4.4.3 Relation between $K_{eq}^{t\to 0}$, $k_a$ and $k_d$

The relation between $K_{eq}^{t\to 0}$, $k_a$ and $k_d$ is:

$$K_{eq}^{t\to 0} = k_a/k_d \tag{13.8B}$$

(see also Eq. (13.8A), in Sub-section 4.2, above).

Thus, if one knows both $K_{eq}^{t\to 0}$ and $k_a^{(mic)}$, one can find the value for $k_d^{(mic)}$, using Eq. (13.8B). For instance in the case of adsorption of HSA onto monosized silica spheres, $K_{eq}^{t\to 0} = 2.78$ L/M and $k_a^{(mic)} = 2.66$ L/M s, so that using Eq. (13.8B), $k_d^{(mic)} = 2.66/2.78 = 0.96$/s; see Docoslis, Wu et al. (2001, 2002).

### 4.4.4 Importance of $k_d$ as the most widely varying kinetic rate constant

For the reasons already mentioned in Sub-section 4.2, above, the most accurate method for deriving the value of $k_d$ (here of $k_d^{(mic)}$), is *not* direct measurement (on account of unavoidable large amounts of hysteresis accompanying such measurements), but rather the use of Eq. (13.8B), above, because both $K_{eq}$ and $k_a$ *can* be measured experimentally with a fair degree of accuracy, provided one measures $K_{eq}^{t\to 0}$ and one measures $k_a$ in less than one second and at very low protein concentrations (Docoslis, Wu et al., 1999).

As is clear from Eq. (13.8B), any variability in the equilibrium adsorption constant, $K_{eq}^{t\rightarrow 0}$, pertaining to different systems involving a protein adsorbing onto a solid substratum, immersed in water, is a consequence of either the variability of $k_a$, or of $k_d$, when changing either the protein or the substratum. In all cases we studied, the pronounced variability between different systems was always due to the variability in $k_d$; see Sub-section 4.4.1, above and see also Absolom and van Oss (1986). Now, it is really to be expected that for a given adsorbing protein (here HSA), dissolved in water, the $k_a$ values should not show much variability because $k_a$ is mainly proportional to the diffusion coefficient (D) of the protein in question, which is not a variable; see Eq. (13.4).

For instance, using aqueous HSA solutions, the $k_a^{(mic)}$ values pertaining to the adsorption of HSA onto polysized silica particles, monosized silica spheres, the same monosized silica spheres that are carboxylated, and talc, were as follows: 2.37; 2.66; 1.61 and 2.20 (all in $10^8$ L/M s). On the other hand, *the* $k_d^{(mic)}$ *values* for the same four systems were: 5.77; 0.96; 38.33; and 0.23 (also all in $10^{-6}\ s^{-1}$, Docoslis, Wu et al., 2001, 2002). Here the extremes of the $k_a^{(mic)}$ values only varied 1.65-fold, whilst the extremes of the $k_d^{(mic)}$ values varied a hundred times more, i.e., 167-*fold*. One may therefore state that what differentiates kinetically between the outcomes of the adsorption of a given dissolved protein onto different substrata is not the kinetics of *arrival* of the protein molecules at the surface of their substrata, but rather the velocity with which each substratum *releases* the just arrived protein molecules. The equilibrium adsorption constant ($K_{eq}^{t\rightarrow 0}$) is therefore energetically the strongest when the velocity with which the just arrived molecules are released from captivity is the slowest.

### 4.4.5 On the lack of merit of expressing equilibrium interaction constants in terms of "$K_d$"

In the more recent literature one increasingly encounters a novel but regrettable habit by authors to express equilibrium interaction constants as "$K_d$" (i.e., *dissociation* equilibrium constants, expressed in units of M/L), without actually having measured $K_d$, but where $K_d$ was obtained via:

$$K_d = 1/K_a \tag{13.17}$$

and where $K_a$ then is expected to be comparable to the $K_{eq}^{t\rightarrow 0}$ term used earlier in this section. Now, it is not possible to measure equilibrium *dissociation* constants without incurring vast errors due to the large amount of hysteresis inherent in the unavoidable pre-adsorption step which is a pre-requisite to all dissociation measurements. It is not certain, however, that many of the authors who are using "$K_d$" actually measured it. It seems equally possible that many of them just measured "$K_a$" and then calculated its inverse ("$K_d$") using Eq. (13.17), above, which may have the advantage that one can express it in terms of M/L, instead of the more unwieldy sounding L/M. The use of "$K_d$" risks in most cases to lead to large inaccuracies and should be avoided. This is even more important in those cases where "$K_d$" has actually been measured, because in such cases its measured value is bound to be exceedingly flawed by reason of the almost insurmountable difficulties in correcting for the large amounts of hysteresis inherent in the procedure (see Section 3, above).

CHAPTER FOURTEEN

# Specific Interactions in Water

## Contents



## 1. Innate and Adaptive Ligand–Receptor Interactions in Biological Systems

Of all the specific ligand–receptor interactions occurring in biological systems (and therefore in aqueous media), those occurring in defense against foreign invading entities in vertebrate organisms are among the best known. Their study is generally categorized under the name *Immunology*. Immunology, i.e., the discipline devoted to the defense of the host organism against foreign (i.e., non-self) intruding entities recognizes two main categories of defensive agents: *innate* and *adaptive* immunological agents.

*Innate* immunological agents are specific ligand-carrying peptides, proteins or cells with which the organism *is born*. In other words, they exist ready-made, from the moment of birth on. Animals, plants and even microbes are endowed with various innate defense molecules.

*Adaptive* immunological agents are specific ligand-carrying proteins or cells which the vertebrate organism *synthesizes*, tailor-made after the first encounter with an antigenic receptor-carrying foreign (often infectious) invading microorganism.

### 1.1 Specific *innate* ligand–receptor interactions

Specific innate ligand–receptor interactions have the advantage that the defense against novel invading microorganisms starts immediately by means of the pre-existing defense molecules, without the necessity for waiting at least five or six days for the production of suitable new defense molecules. The drawback is that their array of useful specificities is fairly limited, to about $10^3$ per host organism.

*Interface Science and Technology*, Volume 16
ISSN 1573-4285, doi: 10.1016/S1573-4285(08)00214-7

### 1.1.1 Small cationic antimicrobial peptides

The *small cationic antimicrobial peptide, lysozyme*, discovered by Sir Alexander Fleming (1920), had for about seven decades remained the first curious, solitary example of what more recently (since about the 1990's) turned out to be a member of the class comprising many hundreds, and possibly close to a thousand different small cationic peptides that are known today; see, e.g., Devine and Hancock (2004). Like many of these more recently discovered small cationic antimicrobial peptides, lysozyme appears to be omnipresent; it is secreted by many different human tissues and cells (e.g., various types of leukocytes), by most vertebrate and invertebrate animals, as well as by many plants, molds and bacteria. Lysozyme's antibacterial activity is exerted by enzymatically lysing bacterial cell walls, but it also has antiviral powers (Merck Index, 11th ed., 1989).

The hundreds of different cationic antimicrobial peptides that have been identified to date tend to have a rather low molecular mass of only a few kD and they are positively charged. These two salient properties which all peptides of this class have in common must have important functional advantages, e.g., in allowing them to thrive in an almost universally negatively charged biological environment. One might fear, for instance, that such cationic peptides would be in constant danger of adhering to almost any biological surface in their vicinity. However, two alleviating conditions tend to prevent such an occurrence.

The first alleviating condition is that although the long-range electrostatic attraction of these cationic peptides to anionic surfaces is attenuated at the fairly elevated ionic strength of most major mammalian extracellular liquid media, i.e., at $\mu \approx 0.15$, it still does allow these peptides to exert, *inter alia*, their specific antiviral activity in lower ionic strength media, such as intracellular liquids. (See, e.g., Devine and Hancock (2004; pp. 1–3).)

The second alleviating condition is that despite the fact that the small size of these peptides causes them to undergo a significant amount of Brownian motion, providing them with an extra repulsive energy of $+1.5$ kT to be added to the hydrophilic repulsion energy caused by their own and their neighbors' hydration pressure, at low ionic strengths the concomitantly longer range electrostatic attraction of these cationic peptides to negatively charged invading microbes is enhanced, which favors their subsequent docking with these microbes. The final bonds between these small peptides and the receptors on their bacterial prey probably are partly electrostatic and partly hydrophobic: In the amino acid sequences of eleven such cationic peptides, published by Brogden et al. (2004, Table 8.1) most amino acids that are not cationic are hydrophobic.

The ultimate aim of the encounter between small cationic peptides and invading microbes is their destruction (e.g., through an enzymatic attack on bacterial cell walls), or simply via their tagging, thus triggering chemotactic signals that call phagocytic or other killer cells to the scene, resulting in the destruction of the invading microbes (see Sub-section 1.1.3, below and see also Devine and Hancock, 2004).

### 1.1.2 Lectin–carbohydrate interactions

Another category of innate, specific defense polymers, produced by plants as well as by various animals, is that of *lectins*, which are proteins or more often, glycoproteins, which react specifically with the *sugar* moieties of glycoproteins on the surfaces of bacteria, viruses, as well as mammalian cells, such as sperm cells and peripheral blood cells.

There are essentially only four different categories of sugars recognized by plant lectins, which are: (I) L-fucose; (II) *N*-acetyl-D-galactosamine and D-galactose; (III) D-mannose and D-glucose; (IV) *N*-acetyl glucosamine; see Pusztai (1991). Descriptions of a few of the more important plant lectins follow below.

#### *A. Lectin specificities directed toward human red cell blood groups and other human blood cells*

I. *Antihuman red cell blood group lectins.* It is a much-studied curiosity that various plant lectins display specificities toward human red cell blood groups, such as A, B and H (H is the antigenic substance of blood group O). Antiblood group H lectins [with antifucose specificity; see Cunningham (1994) for a description of the blood group H epitope] are found in *Ulex europaeus* and *Lotus tetragonolobus* (Pusztai, 1991, p. 30; and Grundbacher, 1973). Some of these lectins are used for the identification of blood groups of units of blood cells to be used in blood transfusion.

Antiblood group A is made, *inter alia*, by *Dolichos biflorus* (with anti-*N*-acetyl-D-galactosamine specificity; see Chattoraj, 1967). Antiblood group B is made by *Sapora japonica* (also with *N*-acetyl-D-galactosamine specificity and with an also-present anti-A specificity which can, however, be removed by the addition of privine-HCl; Chattoraj and Boyd, 1966; see also Cunningham, 1994).

The biological role of the different human blood groups has long been the subject of much speculation (see, e.g., a disquisition on this subject by Garratty, 1994). The physical-biochemical function of the blood group sugars, which are distally placed on the red blood cells' glycocalices, is quite clear: they serve to help avoid intercellular contact while the cells are in the peripheral blood circulation, by means of the strong mutual hydrophilic repulsion between the oligomeric sugars at the cells' surfaces, which repulsion is even stronger than between most proteins, when immersed in aqueous media; see van Oss (2006, Chapter XVIII). There thus is a strong advantage in having distally placed sugars on the glycocalices of peripheral blood cells. However, the question which distally placed sugars would be the most effective in this role is difficult to decide and a determination as to which ones among the three or four possible carbohydrate candidates would be the most hydrophilic (and thus the most strongly repulsive in blood) has never been done. It is known that blood groups A and O among white populations are almost equally represented (43 to 47% of white populations have blood group O and 38 to 41% have blood group A). On the other hand, virtually unmixed populations, such as some native American tribes, have up to 99% blood group O, very little A and no B or AB; see Mohn (1979). It would therefore appear that the various single human blood groups developed independently at least several thousands of years ago in different isolated human tribes, whilst the currently existing variegated distributions of the major human blood groups originally developed in the same manner and later became diluted as a consequence of intermingling with different tribes or populations. The subsequent modern distribution of the major human O, A, B and AB blood groups, as well as various minor ones, also was possibly somewhat influenced by epidemics caused by different infectious microorganisms, against which individuals with given blood groups were more (or less) resistant, as the case may have been; see, e.g., Garratty (1994).

II. *A lectin with antihuman erythrocyte and antihuman lymphocyte specificity*

A phytohemagglutinin (PHA) from kidney bean (*Phaseolus vulgaris*) seeds is a plant lectin which in some varieties of *P. vulgaris* comprises two different molecules, of which one (E) hemagglutinates human erythrocytes and the other (L) hemagglutinates human lymphocytes. The general molecular specificities of both E and L-type lectins are held to be of the anti-*N*-acetyl-D-galactosamine class; see Pusztai (1991, pp. 24–30).

***B. Specificities of concanavalin A*** *Concanavalin A* (Con A) is one of the most-studied plant lectins (see, e.g., Goldstein et al., 1965). Extracted from Jack bean (*Canavalis ensiformis*) meal, it is one of the few lectins that is a pure protein and not, like the majority of lectins, a glycoprotein (Galbraith and Goldstein, 1972).

Con A manifests various antiglucide specificities. It binds *dextran*, which is a polymer of maltose, which is a dimer of glucose (Goldstein et al., 1965, So and Goldstein, 1967). Con A interacts with virus-infected cells (Poste, 1975). Con A also has been shown to act as an immuno-suppressant in mice (Markowitz et al., 1968). Con A's general antisugar specificity is against D-mannose/D-glucose, i.e., group III (see the beginning of Sub-section 1.1.2, above).

***C. Some reviews on plant lectins*** One of the earliest reviews on *plant lectins* was published by Boyd (1963), going back as far as 1888 in his review. Sharon wrote an early review in 1977 and co-published a more extensive one (Sharon and Lis) in 1990. Pusztai (1991) devoted a book to Plant Lectins and a more concise review was published by Pereira (1994). A book solely devoted to Concanavalin A was edited by Chowdhury and Weiss (1975). Finally a Ciba Foundation Symposium (1989, #145) was published on Carbohydrate Recognition in Cellular Function, comprising 14 papers on various aspects of oligosaccharides as recognition structures.

### 1.1.3 Phagocytic interactions

*Human phagocytic cells*, i.e., polymorphonuclear cells (PMN's) also called granulocytes (of which the neutrophils are the most important), as well as monocytes and macrophages[1]), are unable to recognize invading microbes by themselves, but they are capable of approaching, ingesting and subsequently destroying such microbes with the help of various (innate) recognition cytokines. Furthermore, recognition

[1] Neutrophils have a half-life of only a few days; they are mainly found in the peripheral blood circulation, but they are capable of leaving it in response to chemotactic signaling of the presence of invading bacteria in various tissues. Monocytes are young macrophages, of which the latter are much longer lived than neutrophils and are found in lymphnodes, spleen and other tissues; the younger monocytes are still largely to be found in the peripheral blood circulation. Neutrophils are the first line of defense and the monocyte/macrophage system form a close second, much helped by manifold innate recognition and effector proteins and peptides.

molecules such as immunoglobulins, whose origin may be mainly non-specific (and innate), or specific (and adaptive), also play a crucial role in helping to attract phagocytic cells to their prey.

However, on balance, phagocytic cells are here preferentially classified as agents of the innate immune defense system insofar which opsonizes bacteria or other cells, where "opsonization" comes from the Greek το οψον (i.e., a spice, or condiment) which here means "to make palatable." The (mainly non-specific) complement system, or complement "cascade" plays an important role in the opsonization of previously tagged microorganisms. Antigen–antibody binding can trigger the complement cascade into action[2]), but this can also be effected via the aspecific binding of human IgG or IgG3 class antibody onto relatively hydrophobic bacteria. This was first observed with just pooled normal human IgG (also called Cohn's Fraction II, often indicated as F-II), which could of course still be due to the gratuitous presence of actual IgG-class antibodies to one or two bacterial species present in the F-II fraction (van Oss et al., 1975). However, Absolom, van Oss et al. (1982) showed that monoclonal IgG3 (i.e., an IgG3 class monoclonal immunoglobulin from a multiple myeloma patient, of only one single specificity, which specificity remained however unknown) could also opsonize various bacteria, namely: *Escherichia coli, Staphylococcus aureus, Staphylococcus epidermidis and Listeria monocytogenes*, (i.e., one Gram-negative and three Gram-positive species) against all four of which no single monoclonal IgG3 preparation could have cross-reacting specificity so that this monoclonal IgG3 may be confidently assumed to be non-specifically directed against at least three of these four different bacterial species and most likely against all of the four species. In any case, this non-specific monoclonal human IgG3 physically adsorbed onto and opsonized *all bacteria* of the four different bacterial species while they were suspended in just Hanks' balanced physiological salt solution which contained no complement. This resulted in all four cases in significantly enhanced phagocytic engulfment by normal human neutrophils in *in vitro* experiments (Absolom, van Oss et al., 1982). Thus, human monoclonal IgG3 can non-specifically opsonize different bacteria all by itself, by simple aspecific physical adsorption onto the bacterial particles.

### 1.1.4 Natural killer (NK) cells

Natural killer (NK) cells are large T lymphocytes with which the mammalian host organism is endowed by birth. They are pre-programmed with a large number of specificities so that one often finds NK cells that not only have receptors for a fair number of the more common invading bacterial cells as well as cells that have been infected by a virus, but there also are NK cells which can recognize "self" cells, thus preventing them from attacking the host organism's own (non-infected) cells. NK cells kill unwanted cells after achieving contact, followed by engulfment

[2] Human serum complement (C) comprises a series of nine proteins. They are numbered from C1 to C9 (following to the order in which they were discovered), but are acting in the order of C142356789, to lyse (e.g., bacterial) cells, after tagging by an antibody (Ab). The various numbered C fractions split into sub-fractions upon activation. Thus, when an antibody binds to an antigenic site (Ag) on a foreign (i.e., non-self) cell or microbe, the completed sequence C142356789 makes a hole in the prey's cell wall, which then causes it to lyse. However, AgAbC1423 tagging suffices to trigger a phagocytic neutrophil to approach, engulf and destroy it. See, e.g., van Oss et al. (1975) and for a treatise on complement, see Morgan (1990).

and subsequent killing of the unwanted cells through the action of the NK cells' cytotoxic granules, in the same manner as cytotoxic T cells (see Murphy et al., 2008, Chapter 2); see also Sub-section 1.2.3, below).

## 1.2 Specific *adaptive* interactions

*Specific adaptive ligand–receptor interactions* consist of the formation of new defense molecules or cells, triggered by their interaction with newly arrived invading antigenic microorganisms which the host organism had not encountered before. A drawback of the adaptive defense approach, compared to the innate one is that the development of each such an individual tailor-made defense measure takes at least four to five days before it is fully operative. However, one enormous advantage of the adaptive human defense system is that its total array of different possible anti-epitope specificities is of the order of $10^6$ (see (Williamson, 1977)), which is a thousand times more than the pre-existing innate system's capacity; see Sub-section 1.1, above.

### 1.2.1 The immunoglobulins

In earlier evolutionary times ancestral vertebrates only had high-valency (i.e., five- to ten-valency) *immunoglobulins* (Ig), as is still apparent in present-day sharks. However in modern times the valency of most Ig classes in vertebrates is *two*, although among many vertebrate species (including humans) one also still finds a certain proportion of *decavalent* immunoglobulin-M (IgM); see Edberg et al. (1971) and Benedict (1979). IgM is often the first Ig isotype which is produced as the primary response, at the first encounter with a new antigen. Especially with a continuing or repeated challenge with the same antigen, the IgM production tends to change into the production of IgG class antibodies of the same specificity. However, as an immune response to *carbohydrate antigens*, the antibody production may not switch Ig class, but often remains of the IgM variety. In humans the most prevalent Ig is IgG, followed by IgA, IgM, IgD and IgE, where the peripheral blood concentration of IgG is about 1.0 to 1.4% (w/v); IgA, 0.2 to 0.3%; IgD, 0.003% and IgE $10^{-3}$ to $10^{-5}$% (van Oss, 1979). The molecular weight of human IgM is about 900,000, of IgG about 150,000, IgA about 155,00, IgD about 185,000 and IgE 187,000 (van Oss, 1979).

*Human IgG* has four sub-classes (IgG1, IgG2, IgG3 an IgG4) with slightly different biological functions. For example, while all IgG's are the only immunoglobulins capable of passing the placenta, in order to furnish the new-born baby with its first array of (ready-made) antibodies produced by its mother, IgG1 has a half-life of 20 days; IgG2, 24 days; IgG3, 7 days and IgG4, 23 days, so that a newborn is protected against most of its first infections during the first few months, after which it has to make its own. IgG1 and IgG3 are the main immunoglobulins involved in the opsonization of, e.g., bacterial cells and they also are the principal ones that aspecifically bind to monocytes and macrophages (van Oss, 1979), as well as to neutrophils (Absolom, van Oss et al., 1982). Figure 14.1 shows a schematic presentation of an IgG1 molecule; see also van Oss (1979).

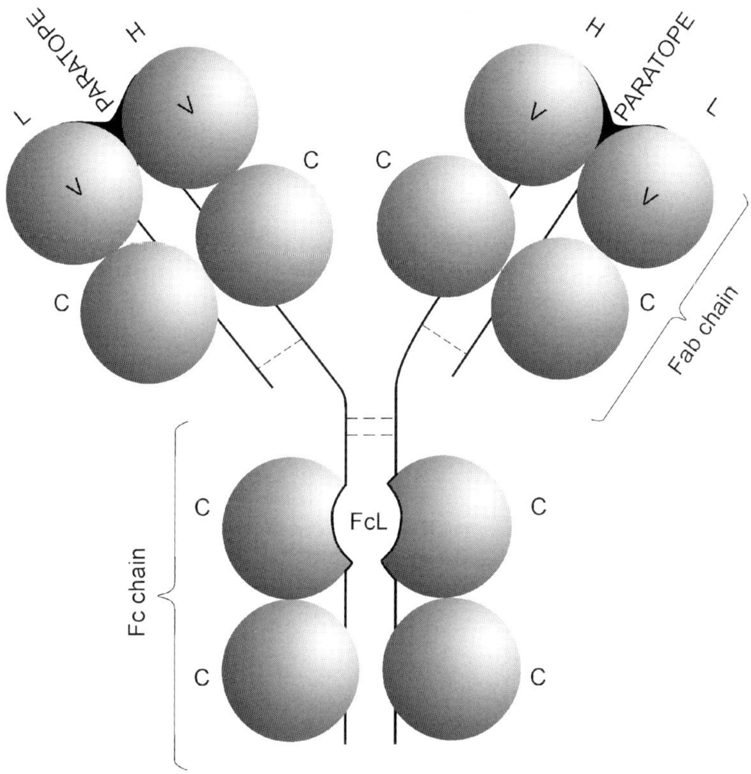


**Figure 14.1** Schematic presentation of some of the salient properties of an IgG1 molecule. The two *heavy* (H) chains have four protein sub-units each and the two *light* (L) chains consist of only two such sub-units each. The two IgG1 H chains are linked together by means of two disulfide bonds, here indicated by two short interrupted lines. Similarly indicated are the single disulfide bonds linking each L chain to its closest H chain. The four *variable* (V) domains are to be found at the amino terminal sides of the complete IgG molecule. When split by means of a papain treatment, just above the two disulfide bonds holding the two H chains together, IgG molecules separate into the two antibody-active fractions, called Fab chains, and one Fc chain, containing the Fc *ligand*, indicated by the abbreviation, FcL. The Fc *ligand* can bind to the first, C1q fraction of the complement cascade (see and it can also strongly bind monocytes as well as to neutrophils, platelets, some lymphocytes, and macrophages, where the latter four bind to the FcL more weakly than to monocytes; see also example II of Sub-section 1.2.1, above. The two antibody-active sites, or *paratopes*, are situated in a concavity or cleft, formed at the amino terminal side, between the two variable (V) moieties, as indicated. Because in the many cases where the paratopic surfaces are at least partly hydrophobic, they have an obvious advantage in being situated in concavities, as this prevents the immunoglobulins from attaching hydrophobically and non-specifically to the multitudinous other proteins and cells that are normally present in their immediate proximity under *in vivo* conditions; see van Oss (1979). It should be recalled that a hydrophobic and a hydrophilic moiety in most cases are quite capable of binding together hydrophobically when encountering each other in an aqueous medium; see Chapter 6, Sub-section 2.1.1 [see also Williamson (1977; Figure 2), for the hydrophobic amino acid content of a number of guinea pig and mouse paratopes of antihapten antibodies].

*Human IgM* consists of five IgG-like proteins arranged in a star-like fashion with the Fc moieties held together at the center of the star by a "joining chain" called "J-chain." As mentioned above, IgM is decavalent, as first determined by Edberg et al. (1971).

*Human IgA* occurs for about 42% in the peripheral blood circulation and for 52% extravascularly, mainly as monomers. However, in the guise of "secretory" IgA (sIgA), IgA occurs mainly as a dimer where in each case two monomers are joined by a "J-chain" of the same type that is also used to hold the five monomeric units of IgM together (see above), whilst secretory IgA has, in addition, a secretory ("s") component to complete the sIgA. *Secretory IgA* is synthesized and secreted by oral, nasal, bronchial and gastro-intestinal tissues in response to local antigenic stimuli by, e.g., viruses such as influenza virus, polio virus, adenovirus and rhinovirus and sIgA is of course tetravalent (van Oss, 1979). IgA (and also sIgA) is the most hydrophilic of all Ig's, as measured via contact angle analysis (see van Oss, 2006, Chapter XVII) and as is also apparent by the fact that it is the only human serum Ig that is difficult to purify by means of reversed-phase liquid chromatography (RPLC), but needs one to resort to hydrophobic interaction chromatography (HIC); see Chapter 6, Sub-section 2.4.1; see also Doellgast and Plaut (1976). Made possible by its extreme hydrophilicity, IgA, by attaching itself to various pathogenic bacteria, endows them with its own surface hydrophilicity so that the immune complexes thus formed can no longer attach to neighboring tissues, thus allowing them to be eliminated in mucous exudates in a natural manner. See also Tomasi (1976); Tomasi and Plaut (1987) and Abraham and Beachy (1987).

*Human IgD* occurs in high concentrations on the surfaces of the B lymphocytes of newborn babies and tends to appear as a very early precursor of the IgM response, formed after a first exposure to a novel antigen (van Oss, 1979).

*Human IgE* occurs at exceedingly tiny concentrations of $\approx 10^{-5}$ to $10^{-4}$% in most people, while this may be up to $10^{-3}$% or more in about 15% of the population, among people who are prone to suffer from atopic allergies (e.g., asthma, hay fever, animal dander allergies, etc.). The evolutionary origin of the IgE isotype was probably to function as an antibody specialized in the defense against parasitic infections (van Oss, 1979).

In general, antibody molecules (Ig's) have two major functions:

I. The first (*adaptive*) one is *specific recognition of* and *binding to* foreign antigenic sites (*epitopes*) on invading microorganisms (in the *adaptive* mode) by means of the Ig's antibody-active sites (*paratopes*) which are situated on the N-terminals of the Fab chains, involving both variable domains on each of the two sides of the Ig molecule; see Figure 14.1.

II. The second (*innate*) function of the Ig's (especially those of the IgG's) is to bind with their Fc chain to various (also *innate*) effector moieties. This Fc chain's ligand comprises parts of both heavy chains' domains; see the schematic sketch shown in Figure 14.1. The Fc moiety binds to receptors on B and T cells (including NK cells), neutrophils, monocytes and macrophages). Equally important is that it binds to complement via the latter's first fraction of its first component, i.e., to C1q, which becomes activated for binding to the Fc fractions of IgG1, IgG2, IgG3, or IgM class Ig's, when for triggering complement binding at least two closely spaced IgG1, IgG2 or IgG3 molecules (having themselves been bound to an antigen) are needed, or just one IgM molecule, bound to an antigen. (IgG3's Fc fraction binds the strongest to complement, the Fc of IgG2 the weakest and that of IgG4 not at all). [When activated by circulating immune (Ag–Ab) complexes, the third component of complement (C3, in the guise of C3b) can bind to a complement receptor on circulating red blood cells, which can cause damage in cases of some auto-immune syndromes; see Geha and Rosen (2008) and see also the legend to Figure 14.1.]

### 1.2.2 Antigen-active sites (*epitopes*) have to be accessible and therefore tend to be hydrophilic

Antigen-active sites, or *epitopes*, which are the recognition sites on living cells such as bacteria or foreign mammalian cells and also on viruses, against which the mammalian immune system needs to provide a defense must, to be recognized, be accessibly placed on the outside of invading cells or particles. As all small invading molecules, particles or cells are or become immersed in the aqueous media of their mammalian host, they need to have a hydrophilic surface, in order to avoid clumping together or to bind onto the surfaces of their host organism.

When the antigen is a protein molecule, its most prominent recognizable moieties are the "elbows" of the protein's tertiary structure. With whale myoglobin [the protein molecule whose tertiary structure was the first one that was completely elucidated, by Kendrew et al. (1960)] the amino acid composition at the "elbows" is readily verified, see, e.g., Kabat (1976, pp. 139, 140), who shows the three-dimensional structure of whale myoglobin, with all amino acids numbered, as well as a numbered list of the amino acids forming the complete primary sequence of the myoglobin molecule.

The most exposed distally placed epitopes of biopolymers consist either of amino acids or of glucides, where the latter usually form the most hydrophilic epitopes.

### 1.2.3 Specific interactions by human lymphocytes

The two major types of human lymphocytes, i.e., B and T lymphocytes play different roles in the specific adaptive immune defense.

*B cells* are the lymphocytes which produce immunoglobulins and which display a number of these on their surfaces. After encountering a novel epitope, a corresponding B cell achieves contact with it, becomes activated and starts producing antibodies which are specific to the recognized epitope, whilst the B cell in question also commences to multiply, leading to many more B cells of the same specificity, thus producing more of the same antibody molecules. Following the continuing presence, or the renewed appearance of the same epitope, which can be shortly after its first appearance or even years later, these specific B cells will again multiply, as a secondary or memory response, which is done much more quickly than was the case at the initial, primary response. The primary and secondary B cell responses together represent the *humoral* immunological defense system, in which the B cells play a crucial but *indirect role*, insofar that they themselves do not function by directly attacking the cells, particles or molecules which carry the novel epitopes, but leave the tagging as well as the attacking to the soluble antibodies they produced.

*T lymphocytes* on the other hand are endowed with specific antigen receptors which are different in detail from, but whose structure bears a strong resemblance to, IgG molecules (see Figure 14.1). T cell receptors (TCR) can select and then cause the whole T cell to approach foreign invading cells, specifically aiming at peptide epitopes placed at the surface of said invading cells, thus ultimately achieving their destruction. Such cytotoxic T cells kill other foreign infectious cells as well as the host organism's own ("self") cells, which have been infected by viruses and after having made contact with the invading cells by means of their TCR, engulf them and finally destroy them with the help of the T cell's intracellular cytotoxic granules (see Murphy et al., 2008; Chapters 2 and 3).

# 2. The Forces Involved in Epitope–Paratope Interactions

## 2.1 Mechanisms and outcomes of epitope–paratope interactions

### 2.1.1 Number of different available paratopes per host organism

Roughly a million possible different paratope compositions have been assumed to be available in an adult host (see, e.g., Williamson, 1977; see also the preamble of Sub-section 1.2, above). Taking an average paratope to comprise only about six amino acids (Kabat, 1976), then, out of a total of about 18 to 20 amino acids to choose from, the above number of approximately one million different paratopic compositions still represents only about 3% of the total of all theoretically possible combinations within the above system, which leads one to hypothesize that not all amino acid combinations, among a group of about six, are chemically favored.

### 2.1.2 Size of the paratope's concavity

As seen above, the paratope concavity *sensu stricto* is composed of hypervariable combinations of, on average, about six amino acids. (The word *hypervariable* is a Greek–Latin hybrid, the use of which is frowned upon by classical scholars. However this term is so widely used in the fields of immunology and immunochemistry that I shall continue to use it, for the sake of clarity.) The paratopic concavity or cleft, comprising this *hypervariable* amino acid sequence, consisting of about six amino acids, is surrounded by two much larger protein domains, together comprising a further (roughly) 280 amino acids which make up the two simply *variable* chains of a complete Fab moiety; see Figure 14.1.

### 2.1.3 Importance of the best three-dimensional fit between epitope and paratope

Both Pressman and Grossberg (1973) and Kabat (1976) stress the importance of as perfect as possible a *three-dimensional fit* of a convex epitope inside the paratopic concavity, for achieving the optimal binding energy between epitope and paratope. Pressman and Grossberg (1973, Chapter 7) also stress the importance of the role of water of hydration in obtaining a perfect fit, but did not consider the effect of squeezing out water of hydration as a function of time; see also Sub-sections 2.1.4 and 2.1.5, below.

### 2.1.4 Minimum distance between the surfaces of epitopes and paratopes

The smallest distance between a concave paratope and its best fitting convex epitope is of course the minimum equilibrium distance between two non-covalently attracting surfaces, $d_0$, which is approximately $0.157 \pm 0.01$ nm and at which minimum distance the non-covalent attractive forces are at a maximum (see Chapter 3, Sub-section 1.1; see also van Oss, 2006, Chapter III). As already indicated in the previous Sub-section 2.1.3, above, the *best fit* between epitope and paratope is equally crucial, because that condition yields the greatest possible *common surface area* between epitope and paratope at which the condition of $d_0 \approx 0.157$ nm can be attained, thus achieving the strongest possible free energy of attraction (see Chapter 3, Tables 3.1, 3.2 and 3.3), regardless which of the three main non-covalent binding forces is the dominant one in any given epitope–paratope interaction.

### 2.1.5 Role of hysteresis

Although the possible role of the epitope's hydration has been amply discussed, e.g., by Pressman and Grossberg (1973), little attention has been given to what happens when some or all of the epitope's bound surface water gets extruded from the paratope's concavity a short time after the epitope–paratope bond was initiated. Now, what happens concomitantly with such an extrusion of water hydration from the epitope's surface, is a significant increase in bond strength. Chapter 13, Sub-sections 3.2 and 3.3 treats the influence on the bond between epitope and paratope when water of hydration is squeezed out, showing that the negative value of $\Delta G_{1w2}$ increases when, subsequent to such a loss of water: $\Delta G_{1w2} \rightarrow \Delta G_{12}$. In Chapter 13 this is treated as part of the onset of *hysteresis*, a phenomenon which is very difficult to avoid, except, curiously, in cases of exclusively electrostatic (EL) interactions, also alluded to in Chapter 3, Sub-section 3.4 and further treated in the present chapter, Sub-section 2.2.3, below.

One must agree again with Pressman and Grossberg (1973) that hydration almost invariably plays a role in the binding between epitopes and paratopes, but its main effect is not what these authors assumed. The main effect is that water of hydration and its extrusion after the bond formation between two surfaces (when immersed in water) is one of the main causes of *hysteresis* which usually starts within a fraction of a second after the beginning of bond formation. It is not easy to measure equilibrium association constants ($K_a$) within, say, the first one tenth of one second but the problem can be largely circumvented by determining $K_a^{t \rightarrow 0}$; see Chapter 13, Sub-sections 3.5 and 3.6. However, even though very early $K_a$ values can be obtained via measurements of $K_a^{t \rightarrow 0}$, this is very rarely done, so that one must fear that virtually all published data on epitope–paratope association constants as well as free energies of association derived from these, practically always have been obtained after significant *hysteresis* had set in. (In actual practice one finds that many workers prefer to measure a *dissociation* constant, $K_d$, which is, however, even more prone to be strongly influenced by hysteresis.)

If one actually were to measure the binding constant of the process ($K_a$) a very short time after its inception, one would still have a reasonably valid binding constant for a given system that could be legitimately comparable to what one might define as a constant, by using the $K_a^{t\to 0}$ approach (see Chapter 13, Sub-section 3.7). Now, binding constants pertaining to immunological reactions may be taken to be rather closely comparable to the desorption of dissolved proteins onto, e.g., silica surfaces, immersed in water. In these cases, it has been determined that the amount of protein which can be desorbed per unit surface area continues to decrease significantly as a function of contact time between protein and substratum. This process was measured after contact time periods from 1 h, up to 24 h, and showed signs of continuing even after that period, albeit at a somewhat diminishing rate, all of which thus being indicative of the continuing strengthening of binding *hysteresis* even after considerable time lapses (Docoslis, Rusinski et al., 2001). It therefore becomes apparent that the binding energy measured (via $K_d$ determinations) for a given Ag–Ab interaction system would not only depend on the nature of the interacting epitopes and paratopes used, but also on the time elapsed between the start of the reaction and the moment its binding energy was determined, via dissociation measurements.

### 2.1.6 Impact of plurivalent binding

There are many occasions where the binding of two, three or more antibody molecules to the same multivalent antigen or cell can trigger reactions which the interaction with only one antibody molecule cannot achieve. A couple of examples are:

I. The activation of phagocytic cells with an antigen having bound at least three antibody molecules in close vicinity of one another (see van Oss et al., 1975, Chapter X).
II. The activation of the first complement factor (C1q) with two or more antibody molecules but not with only one (Morgan, 1990).

The explanation of these is: when n antibody molecules, Ab, interact with one single antigenic (Ag) particle, molecule or cell, this is accounted for in the mass action equation for the equilibrium binding constant ($K_a$), as $[Ab]^n$, thus causing an exponential increase in the value of $K_a$ (in all cases where the molar concentrations are smaller than unity). If the reaction is the following:

$$Ag + n{\cdot}Ab \leftrightarrow AgAb_n \qquad (14.1)$$

then:

$$K_a = [AgAb_n]/[Ag]{\cdot}[Ab]^n \qquad (14.2)$$

[see, e.g., Absolom and van Oss (1986); please note that in Eq. (14.2) square brackets denote the *concentration* of the enclosed entity].

### 2.1.7 Contributions by the two variable Fab domains

In Figure 14.1 it can be seen that in each of the two Fab fractions shown, the (hypervariable) paratopic concavity is partly surrounded by the two variable (V) domains, one on an H chain and one on an L chain. The paratopic concavity itself consists mainly of a rather small *hypervariable* moiety, but its two larger companions surrounding said paratopic concavity, also have a *variable* primary structure, which contribute to the ultimate shape of the *hypervariable* paratopic concavity, whose final structure therefore not only depends on the shape and other properties defined by the sequence of its (roughly) six *hypervariable* amino acids, but is also governed in part by the properties of the two much larger *variable* domains of the Fab fraction that surround them, see Sub-section 2.1.2, above

### 2.1.8 Immune reactions with soluble biopolymer antigens

Soluble antigenic biopolymer molecules (Ag), such as proteins, easily form immune complexes by binding to their corresponding (and also soluble) antibody molecules (Ab). If their binding occurs when Ag and Ab are present in solution at "optimal" or "equivalent" proportions such Ag.Ab complexes tend to be insoluble in aqueous media and thus form *immune precipitates*. However, at conditions of either Ag excess *or* of Ab excess, the Ag·Ab complexes formed remain in solution. Thus, soluble Ag's and Ab's which are complementary to one another are capable of combining in every conceivable proportion, so that there is no true immunochemical stoichiometry in (soluble) Ag–Ab interactions and when interacting at optimal proportions, at which a maximum of immune precipitate is formed, one can only state that they are at their "equivalence point" [see Chapter 7, Sub-section 5.4.3.A and Figures 7.4 and 7.5, above and van Oss (1979) and see also van Oss, Bronson and Absolom (1982) and van Oss, Absolom and Bronson (1982)]. For a discussion of various techniques using immune precipitation in gels, see Chapter 7, Sub-sections 5.4.3 and 5.5.1.

Now, water-soluble biopolymers such as protein Ag's, of a molecular mass greater than about 50 kD (see Chapter 7, Sub-section 3.3, above), as well as their corresponding Ab's, have to repel one another in water to remain soluble. One would therefore have to assume that when a foreign protein Ag of a molecular mass of 50 kD or more were to find itself in the vicinity of a human B lymphocyte while both are immersed in an aqueous liquid medium of a host organism, that protein Ag and the B lymphocyte would repel each other and the dissolved protein Ag would be unable to make contact with that lymphocyte, so that specific Ab formation against said dissolved proteinaceous Ag would not be possible. In actual fact however, nothing would be further from the truth: When both the Ag (the dissolved protein) and the B cell, which happens to bear Ig's endowed with a specific receptor (paratope) for the Ag's epitope, stray into close proximity of one another, they will often be able, at a given favorable orientation of both Ag and the Ab-bearing B cell, to penetrate the mutual repulsion field, achieve contact between epitope and paratope and bind together; see Chapter 13, Sub-section 1.1, above. Indeed, larger, insoluble proteinaceous particles, such as viruses, actually are prone to be broken down enzymatically to smaller, soluble protein sized molecules or peptides, to be processed by the host organism's Ab formation mechanism, where even soluble proteins first become further degraded into small peptides for "presentation" (see, e.g., Adorini and Trembleau, 1994; see also Sub-section 1.2.2, above, for other properties of antigenic proteins).

### 2.1.9 Infrequency of immunogenicity among synthetic polymers

It is well-known that water-soluble, hydrophilic biopolymers such as proteins and glycoproteins are exceedingly immunogenic, when foreign to the immunized host. About 17 years ago we investigated whether Ab's could be elicited (in rabbits) to a number of synthetic polymers of which most were insoluble and a few were soluble, o r even super-soluble in water (e.g., polyethylene glycol). It turned out that very few of the synthetic polymers were immunogenic.

Polyvinyl pyrrolidone (PVP), which is soluble in water, was the only synthetic polymer to which measurable amounts of Ab's could be elicited but with which polymer the highest Ab titers could only be obtained when the highest molecular mass samples were utilized (at ≈360 kD), whilst low titers were obtained with PVP molecules with a molecular mass of 160 and 38 kD and no titers at all at a molecular mass for PVP of 2.5 kD (Naim and van Oss, 1992). Attempts to immunize rabbits with PVP-styrene as well as with PVP-dimethylaminoethyl methacrylate only yielded Ab's to PVP.

No titers at all were obtainable with different varieties of the hydrophobic, water-insoluble polymers: polystyrene, polybutadiene, silicone (oil as well as particles), nor with the more hydrophilic, water-soluble polymers: polyvinyl alcohol, polyvinyl toluene, polyacrylic acid, polymethacrylic acid, polystyrene–sulfonic acid and polyethylene glycol (PEG). Only the strongly water-soluble dextran allows one, with some difficulty, to elicit Ab formation (see, e.g., Edberg et al., 1971, and Kabat, 1976) but with the equally strongly soluble PEG this is virtually impossible. Its pronounced hydrophilicity and its aqueous solubility actually makes it possible to use PEG to induce *tolerance* instead of immunity to protein Ag's, after attaching PEG to each Ag molecule with which one wishes to "tolerize" a host; see Naim and van Oss (1992). The major difference between the immunogenicity, or lack thereof, of dextran and PEG, respectively, lies in the fact that the chemical surface profile of dextran is more "interesting" than that of PEG, because dextran contains more repetitive hexagonal moieties whilst the structure of PEG is only characterized by repetitive oxygen atoms, which actually are the very cause of PEG's strong repulsive properties, in water.

The conclusions one can derive from the above results are that very *hydrophobic* and insoluble polymers, immersed in water, adhere to every cell, particle, macromolecule or biological surface it encounters *in vivo*, which prevents such hydrophobic polymers from attaining access to the appropriate Ab-forming mechanisms. *Hyper-hydrophilic* polymers on the other hand are more likely to cause specific tolerance rather than immunity, especially when they have a fairly bland surface profile. Finally, a high molecular mass tends to have a "steric" advantage in eliciting Ab formation as they are more likely to bind to more of the valencies of IgM molecules still attached to B cells and/or to more IgD or IgG immunoglobulins per B cell, which therefore makes them more likely to activate B lymphocytes bearing complementary Ig's. See, e.g., Edberg et al. (1971), on the binding to more paratopes per IgM or IgG molecule when using Ag's (in this case, dextrans) with the higher molecular masses.

### 2.1.10 Energetics and kinetics of Ag–Ab bond formation and the source of their diversity

As already pointed out when treating the energetics and kinetics of protein adsorption onto metal oxide surfaces, in Chapter 13, the strongest correlation linked to the diversity in bond strengths noted among different adsorption systems, was shown to be the close connection between the *variabilities* in bond strengths and the *kinetics of desorption*, see Chapter 13, Sub-sections 4.4.1 and 4.4.4. More than 20 years ago we reviewed the energetics and kinetics of the formation and dissociation of 14 hapten–antibody systems (Absolom and van Oss, 1986). Haptens are relatively small organic molecules of molecular masses between several hundred Daltons to less than 1 kD. One can only elicit Ab's against such small molecules, by attaching them to a larger molecule (such as bovine serum albumin), functioning as a carrier. Then, upon reacting just the antigenic hapten with the antibody elicited against it, one obtains a very fast kinetic association constant ($k_a$) which is of the order of $10^8$ L/M s and which, with such small organic haptens, varies only between $1.1 \times 10^7$ and $6.2 \times 10^8$ L/M s. The reason for the fast $k_a$ values is the small size of the haptens (in contrast with much heavier biopolymers such as proteins), because, in ideal cases, $k_a$ is directly proportional to the diffusion constant of, in this case, the hapten molecule, see Chapter 13, Eq. (13.4). Now, the equilibrium association constant of the interaction between a hapten and its corresponding antihapten antibody, $K_a$, which is related to the kinetic association rate constant, $k_a$ and the kinetic dissociation rate constant, $k_d$ as:

$$K_a = k_a/k_d. \tag{13.4}$$

Therefore, the extreme variability of the $K_a$ values found for different Ag–Ab as well as for different hapten–Ab systems is not due to the variability of $k_a$, because $k_a$, among haptens of about the same size is not very variable at all. On the other hand, the kinetic dissociation rate constant, $k_d$, is exceedingly variable (see Absolom and van Oss, 1986). One may therefore state that the strong variability of equilibrium association constants, $K_a$, pertaining to Ag–Ab as well as to hapten–Ab systems, is only a consequence of the great variability in the *kinetic dissociation rate constants*, $k_d$. In other words, the variability of the $K_a$ values of Ag–Ab as well as of hapten–Ab systems does not strongly depend upon the *kinetics of attachment* of Ag to Ab or of hapten to Ab, but it is, instead, mainly linked to the difficulty, or the ease, with which the Ag or the hapten manages to *escape* from the paratopic claws of the Ab, after first having become attached to it (see Chapter 13, Sub-section 4.4.1).

*To summarize*, for the construction of a successful synthetic epitope, one needs it to have a high molecular mass, a fair amount of variety, or at least a deviation from strict monotonic linearity in its chemical structure, a reasonable but not excessive degree of aqueous solubility and a composition that is neither extremely hydrophilic nor completely hydrophobic. And, of course, it needs to be foreign (i.e., "non-self") as far as the host organism is concerned. For a few general considerations concerning Ag–Ab interactions, see also Chapter 7, Sub-section 7.2.

## 2.2 Roles of the three non-covalent forces

### 2.2.1 LW forces

The Lifshitz–van der Waals (LW) forces are always active, but may be strongly attenuated when acting on entity (i) while immersed in a liquid with the same, or almost the same Hamaker constant as the immersed materials, which liquid is here taken to be water (w):

$$\Delta G_{iwi}{}^{LW} = -2(\sqrt{\gamma_i{}^{LW}} - \sqrt{\gamma_w{}^{LW}})^2. \qquad (14.3)$$

Equation (14.3) occurs in Chapter 5, as the LW part (term 1) of Eq. (5.1). For the more complex case of the LW free energy when two *different materials*, (1 and 2), are immersed in water (w), see Chapter 6, Eq. (6.1). It should be recalled that $\gamma_i{}^{LW}$ is proportional to the Hamaker constant of material (i); see Chapter 3, Eq. (3.1) for the exact relation between the Hamaker constant (A) and $\gamma_i{}^{LW}$.

With most soluble proteins, $\gamma_i \approx 40$ mJ/m$^2$, while for water, $\gamma_w = 21.8$ mJ/m$^2$, so that using Eq. (14.3), for most proteins their mutual attraction in water is only about $-5.5$ mJ/m$^2$. As most protein molecules here considered repel one another and thus are water-soluble, their contactable surface area ($S_c$) is relatively small, say, of the order of 2.0 nm$^2$, so that, using:

$$\Delta G \text{ (in mJ/m}^2) \cdot S_c/kT = \Delta G \quad \text{(in units kT)}, \qquad (14.4)$$

where $S_c$ is the contactable surface area between two particles or molecules (see Chapter 7), one can estimate the above-mentioned $-5.5$ mJ/m$^2$ as being equivalent to an *attraction* of approximately $-2.7$ kT, which, plus the *repulsion* energy per protein due to Brownian motion of $+1.5$ kT, becomes $-1.2$ kT, which, in water, makes the initial LW attraction between proteins due to LW forces in most cases fairly small.

However, this does not take the generally unavoidable subsequent development of hysteresis into account, especially when much of it takes the form, as is usually the case, of water of hydration being expelled from the paratope–epitope interface, which then causes a change from the initial $\Delta G_{1w2}{}^{LW}$ type of attraction, to the simpler (and stronger) $\Delta G_{12}{}^{LW}$ type; see Chapter 13, Sub-section 3.2. The initial small attractive value for $\Delta G_{1w2}{}^{LW}$ of $-5.5$ mJ/m$^2$ (still taking $S_c$ as only about 2.0 nm$^2$) can then increase up to a value of $\Delta G_{12}{}^{LW} = -59.1$ mJ/m$^2$, which would raise the attractive LW energy between epitope and paratope to about $-29.2$ kT and possibly even higher, as the $S_c$ value is also likely to increase somewhat due to the stronger hysteretic attraction. (The initial attraction giving rise to the expulsion of water needs not only be due to LW forces, but AB and EL forces also usually contribute to this attraction and concomitantly to the expulsion of interstitial water; see below for the AB and EL contributions.) Thus, while the initial LW bond strength between Ag and Ab is usually negligible during the first fraction of a second, after the onset of *hysteresis* caused by the extrusion of water of hydration, the force of the LW bond can turn from negligible to very considerable indeed.

### 2.2.2 AB forces

The Lewis acid–base (AB) interaction energy (in water) comprises, *inter alia*, hydrophobic attraction, which is always active and which, at room temperature, has a value for $\Delta G^{AB}$ of $-102$ mJ/m$^2$, when acting between two completely hydrophobic surfaces. Between two opposing amino acids, where the paratopic one is hydrophobic and the opposing amino acid on the epitope is mildly hydrophilic, the total interaction energy between just these two opposing amino acids would, typically, be about $-20$ mJ/m$^2$. Estimating the contactable surface area ($S_c$) at closest approach between the two opposing amino acids at about 1.0 nm$^2$, one obtains a value for $\Delta G_{1w2}{}^{AB}$ of $\approx -5$ kT, which is significant and still only pertains to two opposing amino acids, in comparison with the total possible interaction between five or six times that number which usually interact between a paratope and an epitope. Thus, even with only one or two paratopic amino acids being hydrophobic, one can already expect a significantly sturdy attraction to ensue by means of hydrophobic interaction forces, which are of the category of *attractive AB forces*. *Repulsive AB forces* on the other hand, serve to keep Ag's and Ab's separate from one another in all cases where they are not complementary (i.e., not attractive to each other on a microscopic level).

### 2.2.3 EL forces

In many cases there is an electrostatic (EL) contribution to paratope–epitope interactions, where for instance one or more acidic amino acids of the paratope attract opposing basic amino acids on the epitope (or vice-versa). However in most cases there may be one or two or even three, e.g., acidic amino acids out of a total of, say, six in a given epitope, which is opposed to basic amino acids from the corresponding paratope. However, a more unusual situation presents itself when the epitope is completely acidic and the paratope uniformly basic. This occurs in cases of auto-antibodies of patients suffering from systemic lupus erythematosus (SLE). This auto-immune disease makes them produce antinuclear (i.e. anti-ds-DNA) auto-antibodies; see, e.g., de Groot et al., 1980; Smeenk et al., 1983; van Oss, Smeenk and Aarden, 1985. In the sera of a number of these patients with anti-ds-DNA antibodies under study, we found that their paratope–epitope interaction systems were exclusively electrostatic and we also found that none of these purely EL Ag–Ab systems underwent hysteresis, that is to say, their Ag–Ab binding energy was the same as the energy required for their dissociation; see also Chapter 13, Sub-section 3.4.

The *reason* for the absence of hysteresis associated with acid DNA/basic anti-DNA interactions in water most likely lies in the fact that, *at closest approach*, the negatively charged epitope (DNA) and the positively charged paratope (anti-DNA) share a single layer of oriented water of hydration, with the water molecules' H atoms attached to the epitopic DNA surface and the water's O atoms attached to the paratopic surface of the anti-DNA antibody, see Figure 14.2. In the illustrated configuration the forces between the oriented monolayer of water molecules and the epitope (DNA) on the one hand and the paratope (anti-DNA) on the other, are both strong enough so that a lateral exclusion of water molecules is not favored from the viewpoint of energetics. This is because in terms of kT units, the energy between tiny water molecules is somewhat smaller than between a water molecule and a larger body which attracts it, so that in cases of electrostatic attraction one single residual layer of water molecules of hydration is the favored configuration.

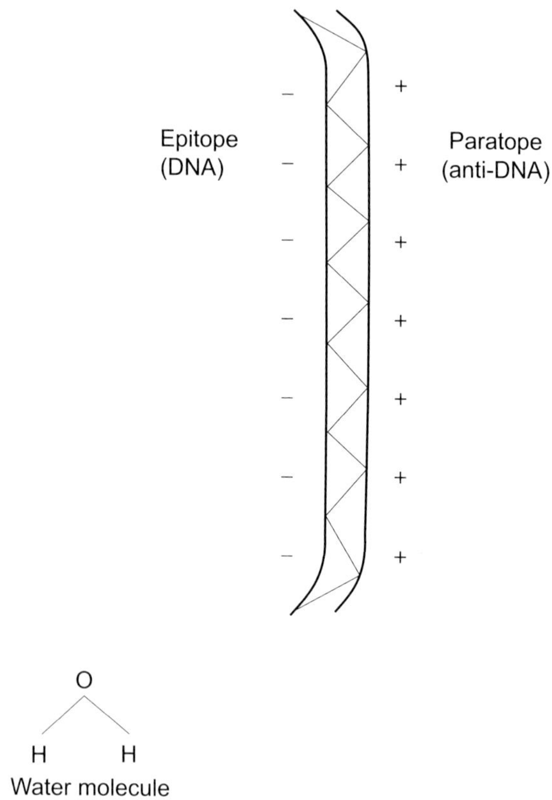


**Figure 14.2** Schematic sketch of the ds-DNA (negatively charged) *epitope* interacting with the (positively charged) anti-ds-DNA *paratope*; see Sub-section 2.2.3, pertaining to EL interactions. Equilibrium is reached when one layer of water of hydration still is present between epitope and paratope, in such a manner that the H atoms of the interstitial water remain attached to the acidic (negatively charged) ds-DNA surface whilst the O atoms of the water remain attached to the basic (positively charged) paratope; see Sub-section 2.2.3 for further explanation.

It should be noted that this behavior prevents the strengthening of bonds through non-electrostatic forces, through the extrusion of water of hydration; see above, Sub-section 2.1.5 and Chapter 13, Sub-section 3.1.

Therefore, in agreement with the experimental observations, because of the lack of expulsion of the last layer of interstitial water between epitope and paratope, as drawn in Figure 14.2, the (EL) free energy of *association* between paratope and epitope in this case has no reason to be different from the energy needed to *dissociate* the bond between these two entities.

## 2.3 Minor or dubious mechanisms of specific bond formation

### 2.3.1 Direct hydrogen-bonding

In various textbooks treating antigen–antibody (Ag–Ab) binding one of the principal mechanisms often proposed is direct hydrogen-bonding between epitope and paratope. Now, on protein surfaces as well as on the surfaces of polysaccharides, in aqueous solution, one finds a large excess of electron-donor sites (i.e., Lewis bases). For instance in Chapter 2, Table 2.3 and also in van Oss (2006, Chapter XVII) it is shown that plasma proteins usually have a $\gamma^+$ less than 1 $mJ/m^2$, and a $\gamma^-$ value of 40 $mJ/m^2$ or more. Also, Docoslis et al. (2000) show that the polysaccharide, dextran, in aqueous solution, has an electron-accepticity ($\gamma^+$) of 2.0 $mJ/m^2$ and an electron-donicity ($\gamma^-$) of 57 $mJ/m^2$; see also van Oss, Chaudhury and Good (1987) on the general high incidence of electron-donating monopolar surfaces. From the known compositions of most epitopes as well as paratopes there is nothing to suggest that these surfaces show any difference in the usual electron-donor/electron-acceptor ratio of about 30/1, in the dissolved state. In addition, direct hydrogen bond formation between an electron-accepting and an electron-donating moiety requires a rather precisely coordinated fit between the two opposing entities, so that for that reason its occurrence is also probably fairly rare. Therefore, direct hydrogen-bonding between epitope and paratope does not seem likely to be a major Ag–Ab binding mechanism.

### 2.3.2 Antigen–antibody binding via "calcium-bridging"

The admixture of $Ca^{2+}$ salts to Ag and Ab can be used to enhance their Ag–Ab complex formation *in vitro*, which effect is usually explained and designated as "calcium-bridging" between Ag molecules (or particles), and Ab molecules. However, as shown in Chapter 8, Sub-section 3.1, what really happens when one adds salts with plurivalent cations to aqueous suspensions of negatively charged particles or to solutions of negatively charged macromolecules, concomitantly with undergoing a decrease in their negative $\zeta$-potentials, these particles' or macromolecules' surface properties change their initial state of hydrophilicity to *hydrophobicity*. Then, from having initially been stably suspended, or dissolved, in water, their newly acquired *hydrophobicity* causes them to agglomerate, or to precipitate; see also the Schulze–Hardy phenomenon (treated in Chapter 8, Sub-section 3.1), which in the hydrophobization of Ag–Ab complexes enhances their insolubilization in water.

# REFERENCES


Abraham, S.N., Beachy, E.H., 1987. In: Gallin, J.I., Fauci, A.S. (Eds.), *Mucosal Immunity*. Raven Press, New York, p. 63.

Abramson, H.A., Moyer, L.S., Gorin, M.H., 1942. *Electrophoresis of Proteins*. Reinhold, New York. Reprinted: Hafner, New York, 1964.

Absolom, D.R., van Oss, C.J., 1986. CRC Crit. Rev. Immunol. 8, 1.

Absolom, D.R., Michaeli, I., van Oss, C.J., 1981a. Electrophoresis 2, 273.

Absolom, D.R., van Oss, C.J., Zingg, W., Neumann, A.W., 1981b. Biochim. Biophys. Acta 670, 74.

Absolom, D.R., van Oss, C.J., Zingg, W., Neumann, A.W., 1982. J. Reticuloendothelial Soc. 31, 59.

Adamson, A.W., 1990. *Physical Chemistry of Surfaces*, 5th Ed. Wiley, New York.

Adorini, L., Trembleau, S., 1994. In: van Oss, C.J., van Regenmortel, M.H.V. (Eds.), *Immunochemistry*. Marcel Dekker, New York, p. 159.

Albertsson, P.-Å., 1986. *Partition of Cell Particles and Macromolecules*. Wiley–Interscience, New York.

Allison, A.C., Humphrey, J.H., 1960. Immunology 3, 95.

Altman, P.L., Ditmar, D.S. (Eds.), 1971. *Blood and Other Body Fluids*. Fed. Am. Soc. Exptl. Biol., Bethesda, MD, p. 12.

Andrade, J.D., Ma, S.M., King, R.N., Gregoris, D.E., 1979. J. Colloid Interface Sci. 49, 291.

Arnold, K., Hermann, A., Gawrisch, K., Pratsch, L., 1988. In: Ohki, S., Flanagan, T.D., Hui, S.W., Mayhew, E. (Eds.), *Molecular Mechanisms of Membrane Fusion*. Plenum Press, New York, p. 255.

Audubert, R., 1955. *Les Ions en Solution, Vol. 1*. Presses Universitaires de France, Paris, p. 141.

Bangham, D.H., Razouk, R.I., 1937. Trans. Faraday Soc. 33, 1459.

Bauer, J. (Ed.), 1994. *Cell Electrophoresis*. CRC Press, Boca Raton, FL.

Becher, P., 1967. In: Schick, M.J. (Ed.), *Nonionic Surfactants*. Marcel Dekker, New York, p. 478.

Bellon-Fontaine, M.-N., Rault, J., van Oss, C.J., 1996. Colloids Surf. B Biointerfaces 7, 47.

Benedict, A.A., 1979. In: Rose, N.R., Milgrom, F., van Oss, C.J. (Eds.), *Principles of Immunology*. Macmillan, New York, p. 14.

Beutner, R., 1913. J. Phys. Chem. 17, 344.

Beutner, R., 1920. In: *Die Entstehung elektrischer Ströme im lebenden Gewebe*. F. Enke (Ed.), Stuttgart.

Bixler, H.J., Michaels, A.S., 1969. Encycl. Polym. Sci. 10, 765.

Boyd, W.C., 1963. Vox Sang. 8, 1.

Bradford, S.C., 1926. In: Alexander, J. (Ed.), *Colloid Science, Vol. 1*. Chemical Catalog. Co., New York, p. 760.

Brogden, K.A., Ackerman, M., Zabner, J., Welsh, M.J., 2004. In: *Mammalian Host Defense Peptides*. Cambridge University Press, Cambridge, UK, p. 189.

Brooks, D.E., 1973. J. Colloid Interface Sci. 43, 687; 700; 714.

Brooks, D.E., Seaman, G.V.F., 1973. J. Colloid Interface Sci. 43, 670.

Bungenberg de Jong, H.G., 1949. In: Kruyt, H.R. (Ed.), *Colloid Science, Vol. II*. Elsevier, Amsterdam, pp. 232, 335, 433.

Busscher, H.J., Stokroos, I., van der Mei, H.C., Rouxhet, P.G., Schakenraad, J.M., 1992. J. Adhes. Sci. Technol. 6, 347.

Casimir, H.B.G., Polder, D., 1948. Phys. Rev. 73, 360.

Cassie, A.B.D., 1948. Discuss. Faraday Soc. 3, 11.

Cassie, A.B.D., Baxter, S., 1944. Trans. Faraday Soc. 40, 546.

Chan, D.Y.C., Mitchell, D.J., Ninham, B.W., Pailthorpe, B.A., 1979. In: Franks, F. (Ed.), *Water, Vol. 6*. Plenum Press, New York, p. 239.

Chattoraj, A., 1967. J. Immunol. 98, 757.



*Interface Science and Technology*, Volume 16
ISSN 1573-4285, doi: 10.1016/S1573-4285(08)00215-9

Chattoraj, A., Boyd, W.C., 1966. J. Immunol. 96, 838.
Chaudhury, M.K., 1984. Short-Range and Long-Range Forces in Colloidal and Macroscopic Systems, Ph.D. Dissertation, State University of New York, Buffalo.
Chaudhury, M.K., 1987. J. Colloid Interface Sci. 119, 174.
Chaudhury, M.K., Good, R.J., 1983. J. Colloid Interface Sci. 94, 292.
Chaudhury, M.K., Good, R.J., 1985. Langmuir 1, 673.
Chowdhury, T.K., Weiss, A.K., 1975. *Concanavalin A*. Plenum Press, New York.
Ciba Foundation Symposium, 1989. *Carbohydrate Recognition in Cellular Function*. Wiley, New York.
Claesson, P.M., 1986. Forces between Surfaces Immersed in Aqueous Solutions, Ph.D. Dissertation, Royal Institute of Technology, Stockholm.
Costanzo, P.M., Giese, R.F., van Oss, C.J., 1990. J. Adhes. Sci. Technol. 4, 267.
Costanzo, P.M., Giese, R.F., van Oss, C.J., 1991. In: Williams, R.A., de Jaeger, N.C. (Eds.), *Advances in Measurement and Control of Colloidal Processes*. Butterworth, London, p. 223.
Costanzo, P.M., Wu, W., Giese, R.F., van Oss, C.J., 1995. Langmuir 11, 1827.
Craig, V.S.J., Ninham, B.W., Pashley, R.M., 1993. J. Phys. Chem. 97, 10192.
Cunningham, R.K., 1994. In: van Oss, C.J., van Regenmortel, M.H.V. (Eds.), *Immunochemistry*. Marcel Dekker, New York, p. 319.
Davidson, D.W., 1973. In: Franks, F. (Ed.), *Water, Vol. 2*. Plenum Press, New York, p. 115.
de Gennes, P.-G., 1990. Rev. Mod. Phys. 57, 827.
de Groot, E.R., Lamers, M.C., Aarden, L.A., Smeenk, R.J.T., van Oss, C.J., 1980. Immunol. Commun. 9, 515.
Debye, P., 1920. Phys. Z. 21, 178.
Debye, P., 1921. Phys. Z. 22, 302.
Derjaguin, B.V., 1934. Kolloid Zh. 69, 155.
Derjaguin, B.V., 1954. Disc. Faraday Soc. 18, 85.
Derjaguin, B.V., Landau, L.D., 1941. Acta Physicochim. URSS 14, 633.
Devine, A., Hancock, R.E.W. (Eds.), 2004. *Mammalian Host Defense Peptides*. Cambridge University Press, Cambridge, UK.
Dobry, A., 1948. Bull. Soc. Chim. Belg. 57, 280.
Docoslis, A., Wu, W., Giese, R.F., van Oss, C.J., 1999. Colloids Surf. B Biointerfaces 13, 83.
Docoslis, A., Giese, R.F., van Oss, C.J., 2000. Colloids Surf. B Biointerfaces 19, 147.
Docoslis, A., Rusinski, L.A., Giese, R.F., van Oss, C.J., 2001. Colloids Surf. B Biointerfaces 22, 267.
Docoslis, A., Wu, W., Giese, R.F., van Oss, C.J., 2001. Colloids Surf. B Biointerfaces 22, 217.
Docoslis, A., Wu, W., Giese, R.F., van Oss, C.J., 2002. Colloids Surf. B Biointerfaces 25, 97.
Doellgast, G.J., Plaut, A.G., 1976. Immunochemistry 13, 135.
Donnan, F.G., 1924. Chem. Rev. 1, 73.
Donnan, F.G., Green, G., 1914. Proc. R. Soc. London 90, 460.
Dupré, A., 1869. *Théorie Mécanique de la Chaleur*. Gauthier–Villars, Paris. p. 367.
Edberg, S.C., Bronson, P.M., van Oss, C.J., 1971. Prep. Biochem. 1, 249.
Eisenberg, D., Kauzmann, W., 1969. *The Structure and Properties of Water*. Oxford University Press, Oxford, pp. 145; 169.
Fleming, A., 1920. Proc. Roy. Soc. London B 93, 306.
Flory, P.J., 1953. *Principles of Polymer Chemistry*. Cornell University Press, Ithaca, NY.
Fowkes, F.M., 1963. J. Phys. Chem. 67, 2538.
Fowkes, F.M., 1964. Ind. Eng. Chem. (Dec.), 40.
Fowkes, F.M., 1965. In: Ross, S. (Ed.), *Chemistry at Interfaces*. Am. Chem. Soc., Washington, DC, p. 1.
Fowkes, F.M., 1967. J. Adhes. Sci. Technol. 1, 7.
Fowkes, F.M., 1968. J. Colloid Interface Sci. 28, 493.
Fowkes, F.M., McCarthy, D.C., Mostafa, M.A., 1980. J. Colloid Interface Sci. 78, 200.
Fowkes, F.M., Jinnai, H., Mostafa, M.A., Anderson, F.W., Moore, R.J., 1982. In: ACS Symposium Series, vol. 200. American Chemical Society, Washington, DC, p. 307.

Galbraith, W., Goldstein, I.J., 1972. Biochemistry 11, 3976.
Garratty, G., 1994. In: Garratty, G. (Ed.), *Immunobiology of Transfusion Medicine*. Marcel Dekker, New York, p. 201.
Geha, R., Rosen, F., 2008. In: *Case Studies in Immunology*. Garland Science/Taylor and Francis, New York, p. 25.
Giese, R.F., van Oss, C.J., 2002. *Colloid and Surface Properties of Clays and Related Materials*. Marcel Dekker, New York.
Giese, R.F., Wu, W., van Oss, C.J., 1996. J. Dispers. Sci. Technol. 17, 527.
Girifalco, L.A., Good, R.J., 1957. J. Phys. Chem. 61, 904.
Goldstein, I.J., Hollerman, C.E., Smith, E.E., 1965. Biochemistry 4, 876.
Good, R.J., 1975. J. Colloid Interface Sci. 52, 308.
Good, R.J., 1979. Surf. Colloid Sci. 11, 1.
Grabar, P., Burtin, P., (Eds.) 1964. Immuno-Electrophoretic Analysis, Elsevier, Amsterdam.
Grabar, P., Williams, C.A., 1953. Biochim. Biophys. Acta 10, 193.
Grabar, P., Williams, C.A., 1955. Biochim. Biophys. Acta 17, 67.
Graciaa, Morel, G., Saulner, P., Schechter, R.S., 1995. J. Colloid Interface Sci. 172, 131.
Grasso, D., Subramaniam, K., Butkus, M., Strevett, K., Bergendahl, P.J., 2002. Rev. Environ. Sci. Biotechnol. 1, 17.
Grundbacher, F.J., 1973. Science 181, 461.
Gutmann, V., 1978. *The Donor–Acceptor Approach to Molecular Interactions*. Plenum Press, New York.
Hamaker, H.C., 1936. Recueil Trav. Chim. Pays-Bas 55, 1015.
Hamaker, H.C., 1937a. Physica 4, 1058.
Hamaker, H.C., 1937b. Recueil Trav. Chim. Pays-Bas 56, 3.
Hamaker, H.C., 1937c. Recueil Trav. Chim. Pays-Bas 56, 727.
Hamilton, W.C., 1974. J. Colloid Interface Sci. 47, 672.
Hancock, R.E.W., Devine, A., 2004. In: Devine, A., Hancock, R.E.W. (Eds.), *Mammalian Host Defense Peptides*. Cambridge University Press, Cambridge, UK, p. 1.
Hardy, W.B., 1900. J. Phys. Chem. 33, 385.
Harkins, W.D., 1952. *The Physical Chemistry of Surface Films*. Reinhold, New York. p. 25.
Hauxwell, F., Ottewill, R.H., 1970. J. Colloid Interface Sci. 34, 473.
He, X.M., Carter, D.C., 1992. Nature 358, 209.
Henry, D.C., 1931. Proc. R. Soc. London A 133, 106.
Hiemenz, P.C., Rajagopalan, R., 1997. *Principles of Colloid and Surface Chemistry*. Marcel Dekker, New York.
Hildebrand, J.H., 1979. Proc. Natl. Acad. Sci. USA 76, 194.
Hirsch-Ayalon, P., 1956. Recueil Trav. Chim. Pays-Bas 75, 1065.
Hirsch-Ayalon, P., 1957. J. Polym. Sci. 23, 697.
Hough, D.B., Thompson, L., 1987. In: Schick, M.J. (Ed.), *Nonionic Surfactants*, 2nd. Ed. Marcel Dekker, New York, p. 601.
Hough, D.B., White, L.R., 1980. Adv. Colloid Interface Sci. 14, 3.
Hückel, E., 1924. Phys. Z. 23, 204.
Hunter, R.J., 1981. *Zeta Potential in Colloid Science*. Academic Press, London.
Israelachvili, J.N., 1973. J. Chem. Soc. Faraday Trans. II 69, 1729.
Israelachvili, J.N., 1974. Quart. Rev. Biophys. 6, 341.
Israelachvili, J.N., 1985. *Intermolecular and Surface Forces*. Academic Press, London.
Israelachvili, J.N., 1991. *Intermolecular and Surface Forces*. Academic Press, London.
Israelachvili, J.N., Pashley, R.M., 1984. J. Colloid Interface Sci. 98, 500.
Israelachvili, J.N., Wennerström, W., 1996. Nature 379, 199.
Janczuk, B., Chibowski, E., Choma, I., Davidowicz, A.L., Bialopiotrowicz, T., 1990. Mater. Chem. Phys. 25, 185.

Jasper, J.J., 1972. J. Phys. Chem. Ref. Data 1 (4), 841.
Kabat, E.A., 1976. *Structural Concepts in Immunology and Immunochemistry*. Holt Rinehart and Winnton, New York.
Kaelble, D.H., 1970. J. Adhes. 2, 66.
Keesom, W.H., 1915. Proc. R. Neth. Acad. Sci. 18, 636.
Keesom, W.H., 1920. Proc. R. Neth. Acad. Sci. 23, 939.
Keesom, W.H., 1921. Phys. Z. 22, 129.
Kendrew, J.C., Dickerson, R.E., Strandberg, B.E., Hart, R.G., Davies, D.R., Phillips, D.C., Shore, V.C., 1960. Nature 185, 422.
Ku, C.A., Henry, J.D., Siriwardane, R., Roberts, L., 1985. J. Colloid Interface Sci. 106, 377.
Labib, M.E., Williams, R., 1984. J. Colloid Interface Sci. 97, 356.
Labib, M.E., Williams, R., 1986. Colloid Polym. Sci. 264, 533.
Labib, M.E., Williams, R., 1987. J. Colloid Interface Sci. 115, 330.
Lange, H., 1967. In: Schick, M.J. (Ed.), *Nonionic Surfactants*. Marcel Dekker, New York, p. 443.
Lewis, G.N., 1923. *Valence and Structure of Atoms and Molecules*. American Chemical Society/The Chemical Catalog Co., New York. [Out of print, but available through UMI/Books in Demand, Ann Arbor, MI.]
Liesegang, R.E., 1896. Naturwissenschaft 11, 353.
Liesegang, R.E., 1898. Chemische Reaktionen in Gallerten, Düsseldorf.
Lifshitz, E.M., 1955. Zh. Eksp. Teor. Fiz. 29, 94.
London, F., 1930. Z. Phys. 63, 245.
Luzar, A., Chandler, D., 1996. Nature 373, 55.
Mancini, G., Carbonara, A.O., Heremans, J.F., 1965. Immunochemistry 2, 235.
Markowitz, H., Person, D.A., Gitnick, G.L., Ritts, R.E., 1968. Science 163, 476.
Meryman, H.T., 1966. In: Meryman, H.T. (Ed.), *Cryobiology*. Academic Press, New York, p. 63.
Meylheuc, T., van Oss, C.J., Bellon-Fontaine, M.-N., 2001. J. Appl. Microbiol. 91, 822.
Michaeli, I., Absolom, D.R., van Oss, C.J., 1980. J. Colloid Interface Sci. 77, 586.
Michaels, A.S., 1976. Pure Appl. Chem. 46, 193.
Mohn, J.F., 1979. In: Rose, N.R., Milgrom, F., van Oss, C.J. (Eds.), *Principles of Immunology*. Macmillan, New York, p. 338.
Morgan, B.P., 1990. *Complement*. Academic Press, New York.
Murphy, K., Travers, P., Walport, M. (Eds.), 2008. *Janeway's Immunobiology*. Garland Science/Taylor and Francis, New York.
Naim, J.O., van Oss, C.J., 1992. Immunol. Invest. 21, 649.
Napper, D.H., 1983. *Polymeric Stabilization of Colloidal Dispersions*. Academic Press, New York.
Nemethy, G., Scheraga, H.A., 1962. J. Chem. Phys. 36, 3382.
Neumann, A.W., Good, R.J., Hope, C.J., Sejpal, M., 1974. J. Colloid Interface Sci. 49, 291.
Onda, T., Shibuichi, S., Satoh, N., Tsujii, K., 1996. Langmuir 12, 2125.
Ottewill, R.H., 1967. In: Schick, M.J. (Ed.), *Nonionic Surfactants*. Marcel Dekker, New York, p. 627.
Ouchterlony, Ö., 1949. Arkiv Kemi 1, 43; 55.
Ouchterlony, Ö., 1968. *Handbook of Immunodiffusion and Immuno-Electrophoresis*. Ann Arbor Science, Ann Arbor, MI.
Overbeek, J.Th.G., 1943. Kolloid Beihefte 54, 287.
Overbeek, J.Th.G., 1952. In: Kruyt, H.R. (Ed.), *Colloid Science, Vol. 1*. Elsevier, Amsterdam.
Overbeek, J.Th.G., 1977. J. Colloid Interface Sci. 58, 408.
Overbeek, J.Th.G., Wiersema, P.H., 1952. In: Bier, M. (Ed.), *Electrophoresis, Vol. II*. Academic Press, New York, p. 1.
Owens, D.K., Wendt, R.C., 1969. J. Appl. Polym. Sci. 13, 1741.
Parsegian, V.A., Fuller, N., Rand, R.P., 1979. Proc. Natl. Acad. Sci. USA 76, 2750.
Pashley, R.M., McGuigan, P.M., Ninham, B.W., Evans, D.F., 1985. Science 229, 1088.

Pereira, M.E.A., 1994. In: van Oss, C.J., van Regenmortel, M.H.V. (Eds.), *Immunochemistry*. Marcel Dekker, New York, p. 259.
Pfeffer, W., 1877. Osmotische Untersuchungen, Leipzig.
Poste, G., 1975. In: Chowdhury, T.K., Weiss, A.K. (Eds.), *Concanavalin A*. Plenum Press, New York, p. 117.
Pressman, D., Grossberg, A.L., 1973. *The Structural Basis of Antibody Specificity*. W.A. Benjamin, Reading, MA.
Pusztai, A., 1991. *Plant Lectins*. Cambridge University Press, Cambridge, UK.
Rao, C.N.R., 1972. In: Franks, F. (Ed.), *Water, Vol. 1*. Plenum Press, New York, p. 93.
Rekveld, S. 1997. Ellipsometric studies of protein adsorption onto hard surface in a flow cell. Ph.D. dissertation. Twente University, p. 30.
Rose, N.R., 1979. In: Rose, N.R., Milgrom, F., van Oss, C.J. (Eds.), *Principles of Immunology*. Macmillan, New York, p. 277.
Rosenberg, M., Gutnick, D., Rosenberg, E., 1980. FEMS Microbiol. Lett. 9, 29.
Schick, M.J. (Ed.), 1965. *Nonionic Surfactants*. Marcel Dekker, New York.
Schick, M.J. (Ed.), 1987. *Nonionic Surfactants*, 2nd. Ed. Marcel Dekker, New York.
Schulze, H., 1882. J. Prakt. Chem. 25, 431.
Schulze, H., 1883. J. Prakt. Chem. 27, 320.
Sharon, N., 1977. Sci. Am. 236 (6), 108.
Sharon, N., Lis, H., 1990. FASEB J. 4, 3198.
Shaw, D.J., 1969. *Electrophoresis*. Academic Press, New York.
Smeenk, R.J.T., Aarden, L.A., van Oss, C.J., 1983. Immunol. Commun. 12, 177.
So, L.L., Goldstein, I.J., 1967. J. Biol. Chem. 242, 1617.
Sober, H. (Ed.), 1968. *Handbook of Biochemistry*. CRC Co., Cleveland, p. C-12.
Spelt, J.K., Absolom, D.R., Zingg, W., van Oss, C.J., Neumann, A.W., 1982. Cell Biophys. 4, 117.
Stephen, H., Stephen, T., 1963. *Solubilities of Organic and Inorganic Compounds, Vol. 1 Part 1*. Pergamon Press, Oxford.
Stern, K.H., 1967. *Bibliography of Liesegang Rings*, 2nd Ed. Miscellaneous Publication, vol. 292. National Bureau of Standards, Washington, DC.
Tabor, D., Winterton, R.H.S., 1969. Proc. R. Soc. London. A 312, 435.
Tomasi, T.B., 1976. *The Immune System of Secretions*. Prentice–Hall, Englewood Cliffs, NJ.
Tomasi, T.B., Plaut, A.G., 1987. In: Gallin, J.I., Fauci, A.S. (Eds.), *Mucosal Immunity*. Raven Press, New York, p. 31.
van Oss, C.J., 1962. Encyclopédie Vétérinaire, vol. 19, 457.
van Oss, C.J., 1968. J. Colloid Interface Sci. 27, 684.
van Oss, C.J., 1979. In: Rose, N.R., Milgrom, F., van Oss, C.J. (Eds.), *Principles of Immunology*. Macmillan, New York, p. 41.
van Oss, C.J., 1984a. In: Atassi, M.Z., van Oss, C.J., Absolom, D.R. (Eds.), *Molecular Immunology*. Marcel Dekker, New York.
van Oss, C.J., 1984b. In: Matijevic, E. (Ed.), *Surface and Colloid Science, Vol. 13*. Plenum Press, New York, p. 115.
van Oss, C.J., 1989a. J. Dispers. Sci. Technol. 9, 561.
van Oss, C.J., 1989b. J. Protein Chem. 8, 661.
van Oss, C.J., 1990. J. Protein Chem. 9, 487.
van Oss, C.J., 1991a. Biofouling 4, 25.
van Oss, C.J., 1991b. Polym. Prepr. 32, 598.
van Oss, C.J., 1994. *Interfacial Forces in Aqueous Media*. Marcel Dekker, New York.
van Oss, C.J., 1998. In: Delves, P.J., Roitt, I.M. (Eds.), *Encyclopedia of Immunology*, 2nd. Ed. Academic Press, London, p. 163.
van Oss, C.J., 1999. Int. J. Biochromatogr. 4, 139.

van Oss, C.J., 2003. J. Mol. Recognit. 16, 177.
van Oss, C.J., 2006. *Interfacial Forces in Aqueous Media*, 2nd Ed. CRC Press/Taylor and Francis, Boca Raton, FL.
van Oss, C.J., Costanzo, P.M., 1992. J. Adhes. Sci. Technol. 6, 477.
van Oss, C.J., Fontaine, M., 1961. Z. f. Immunitätsforschung 122, 45.
van Oss, C.J., Giese, R.F., 1995. Clays Clay Miner. 42, 474.
van Oss, C.J., Giese, R.F., 2004. J. Dispers. Sci. Technol. 25, 631.
van Oss, C.J., Giese, R.F., 2005. J. Adhes. 81, 237.
van Oss, C.J., Good, R.J., 1984. Colloids Surf. 8, 373.
van Oss, C.J., Good, R.J., 1988. J. Protein Chem. 7, 179.
van Oss, C.J., Good, R.J., 1991. J. Dispers. Sci. Technol. 12, 95.
van Oss, C.J., Good, R.J., 1992. Langmuir 8, 2877.
van Oss, C.J., Good, R.J., 1996. J. Dispers. Sci. Technol. 17, 433.
van Oss, C.J., Heck, Y.S.L., 1961. Z. f. Immunitaetsforschung 122, 44.
van Oss, C.J., Hirsch-Ayalon, P., 1959. Science 129, 1365.
van Oss, C.J., Mohn, J.F., 1970. Vox Sang. 19, 432.
van Oss, C.J., Friedmann, J.C., Fontaine, M., Drieux, H., 1960. C. R. Acad. Sci. Paris 250, 4067.
van Oss, C.J., Fike, R.M., Good, R.J., Reinig, J.M., 1974. Anal. Biochem. 60, 242.
van Oss, C.J., Gillman, C.F., Neumann, A.W., 1975. *Phagocytic Engulfment and Cell Adhesiveness*. Marcel Dekker, New York.
van Oss, C.J., Omenyi, S.N., Neumann, A.W., 1979. Colloid Polym. Sci. 257, 737.
van Oss, C.J., Absolom, D.R., Bronson, P.M., 1982. Immunol. Commun. 11, 139.
van Oss, C.J., Bronson, P.M., Absolom, D.R., 1982. Immunol. Commun. 11, 129.
van Oss, C.J., Moore, L.L., Good, R.J., Chaudhury, M.K., 1985. J. Protein Chem. 4, 245.
van Oss, C.J., Smeenk, R.J.T., Aarden, L.A., 1985. Immunol. Invest. 14, 245.
van Oss, C.J., Good, R.J., Chaudhury, M.K., 1986a. J. Protein Chem. 5, 385.
van Oss, C.J., Good, R.J., Chaudhury, M.K., 1986b. J. Colloid Interface Sci. 110, 604.
van Oss, C.J., Chaudhury, M.K., Good, R.J., 1987. Adv. Colloid Interface Sci. 28, 35.
van Oss, C.J., Chaudhury, M.K., Good, R.J., 1988. Chem. Rev. 88, 927.
van Oss, C.J., Chaudhury, M.K., Good, R.J., 1989. Sep. Sci. Technol. 24, 15.
van Oss, C.J., Ju, L., Chaudhury, M.K., Good, R.J., 1989. J. Colloid Interface Sci. 128, 313.
van Oss, C.J., Arnold, K., Good, R.J., Gawrisch, K., Ohki, S., 1990. J. Macromol. Sci. Chem. A 27 (5), 563.
van Oss, C.J., Giese, R.F., Costanzo, P.M., 1990. Clays Clay Miner. 38, 151.
van Oss, C.J., Good, R.J., Busscher, H.J., 1990. J. Dispers. Sci. Technol. 11, 75.
van Oss, C.J., Giese, R.F., Norris, J., 1992. Cell Biophys. 20, 253.
van Oss, C.J., Giese, R.F., Li, Z., Murphy, K., Norris, J., Chaudhury, M.K., Good, R.J., 1992. J. Adhes. Sci. Technol. 6, 413.
van Oss, C.J., Giese, R.F., Wentzek, R., Norris, J., Chaudhury, M.K., Good, R.J., 1992. J. Adhes. Sci. Technol. 6, 503.
van Oss, C.J., Giese, R.F., Wu, W., 1997. J. Adhes. 63, 71.
van Oss, C.J., Giese, R.F., Wu, W., 1998. J. Dispers. Sci. Technol. 19, 1221.
van Oss, C.J., Docoslis, A., Giese, R.F., 2001. Colloids Surf. B Biointerfaces 22, 285.
van Oss, C.J., Giese, R.F., Docoslis, A., 2001. Mol. Cell Biol. 45, 721.
van Oss, C.J., Wu, W., Docoslis, A., Giese, R.F., 2001. Colloids Surf. B Biointerfaces 20, 87.
van Oss, C.J., Docoslis, A., Giese, R.F., 2002. In: Mittal, K.L. (Ed.), *Contact Angle*, In: Wettability and Adhesion, vol. 2. VSP Publ., Zeist, The Netherlands, p. 3.
van Oss, C.J., Giese, R.F., Good, R.J., 2002. J. Dispers. Sci. Technol. 23, 455.
van Oss, C.J., Giese, R.F., Docoslis, A., 2005. J. Dispers. Sci. Technol. 26, 585.
Verwey, E.J.W., Overbeek, J.Th.G., 1948. *Theory of the Stability of Lyophobic Colloids*. Elsevier, Amsterdam.

Visser, J., 1972. Adv. Colloid Interface Sci. 3, 331.
von Smoluchowski, M., 1918. Z. F. Physik. Chem. 92, 129.
von Smoluchowski, M., 1921. In: Grätz, L. (Ed.), Handbuch der Elektrizität und des Magnetismus, Vol. II. Ambrosius Barth Verlag, Leipzig, p. 366.
Washburn, E.W. (Ed.), 1921. *International Critical Tables*. McGraw–Hill, New York, pp. 466–467.
Weast, R.J., 1970/71. *Handbook of Chemistry and Physics*, 51st Ed. Chem. Rubber Co., Cleveland.
White, J.G., 1968. Blood 31, 604.
Williamson, A.R., 1977. In: Glynn, L.E., Steward, M.W. (Eds.), *Immunochemistry*. Wiley, New York, p. 141.
Wu, W., 1994. Linkage Between $\zeta$-Potential and Electron-Donicity of Charged Polar Surfaces—The Mechanisms of Flocculation and Repeptization of Particle Suspensions. Ph.D. Dissertation, State University of New York at Buffalo.
Wu, W., Giese, R.F., van Oss, C.J., 1994a. Colloids Surf. A Physicochem. Eng. Aspects 89, 241.
Wu, W., Giese, R.F., van Oss, C.J., 1994b. Colloids Surf. A Physicochem. Eng. Aspects 89, 253.
Wu, W., Giese, R.F., van Oss, C.J., 1996. Powder Technol. 89, 129.
Young, T., 1805. Philos. Trans. R. Soc. 95, 65.
Zaslavsky, B.Y., 1995. *Aqueous Two-Phase Partitioning*. Marcel Dekker, New York.
Zsigmondy, R., 1901. Z. Anal. Chem. 40, 697.

# Subject Index

**J**



**K**



**L**